FISH PHYSIOLOGY
Recent Advances

Fish physiology is a subject of increasing research interest, both as a developing area within animal physiology generally and as a foundation for applied work in fisheries, aquaculture and pollution ecology. This book reviews a number of major advances in this area including haemopoiesis, acid-base regulation, physiology of the circulation, digestive physiology and physiological toxicology. It will therefore interest all workers in pure and applied fish biology.

FISH PHYSIOLOGY:
Recent Advances

Edited by

STEFAN NILSSON and SUSANNE HOLMGREN

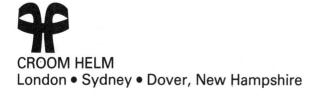

CROOM HELM
London • Sydney • Dover, New Hampshire

© 1986 Stefan Nilsson and Susanne Holmgren
Croom Helm Ltd, Provident House, Burrell Row,
Beckenham, Kent BR3 1AT
Croom Helm Australia Pty Ltd, Suite 4, 6th Floor
64-76 Kippax Street, Surry Hills, NSW 2010, Australia

British Library Cataloguing in Publication Data
Fish physiology: recent advances.
1. Fishes—Physiology
I. Nilsson, Stefan   II. Holmgren, Susanne
   597′.01   QL639.1
ISBN 0-7099-1837-2

Croom Helm, 51 Washington Street, Dover,
New Hampshire 03820, USA

Library of Congress Cataloging-in-Publication Data
Fish physiology.

Includes index.
1. Fishes — Physiology. I. Nilsson, Stefan,
1946-   . II. Holmgren, Susanne, 1946-
QL639.1.F574 1986   597′.01   86-13471
ISBN 0-7099-1837-2

51, 505

Typeset in 10pt Times Roman by Leaper & Gard Ltd, Bristol, England

Printed and bound in Great Britain by
Biddles Ltd, Guildford and King's Lynn

CONTENTS

Preface

List of Contributors

1. Physiology of Haemopoiesis
   Ragnar Fänge   1

2. Mechanisms and Limitations of Fish Acid-Base Regulation
   Norbert Heisler   24

3. Physiological Investigations of Marlin
   Charles Daxboeck and Peter S. Davie   50

4. Fish Cardiology: Structural, Haemodynamic, Electromechanical and Metabolic Aspects
   Kjell Johansen and Hans Gesser   71

5. Control of Gill Blood Flow
   Stefan Nilsson   86

6. Exercise
   Pat J. Butler   102

7. Gastro-intestinal Peptides in Fish
   Susanne Holmgren, Ann-Cathrine Jönsson and Björn Holstein   119

8. Gastro-intestinal Physiology: Rates of Food Processing in Fish
   David J. Grove   140

9. Filtration in the Perfused Hagfish Glomerulus
   Jay A. Riegel   153

10. Physiological Methods in Fish Toxicology: Laboratory and Field Studies
    Lars Förlin, Carl Haux, Tommy Andersson, Per-Erik Olsson and Åke Larsson   158

11. Toxicity Testing Procedures
    Göran Dave   170

Index   196

This book is dedicated to Professor Emeritus Ragnar Fänge

PREFACE

Fishes are very successful vertebrates and have adapted to a wide range of environmental conditions, from the deep ocean to the smallest brook or pond. The physiological background to these environmental adaptations is, obviously, far from clear, and provides fish physiologists with many challenges. The number of extant fish species has been estimated to be in excess of 20 000, and only relatively few of these have been subject to physiological studies. Yet among these animals can be found many physiological systems different from those of the land-dwelling vertebrates, and also systems similar to those of the 'higher' vertebrates but at a different level of phylogenetic development.

Apart from the rapidly increasing interest in basic fish physiology, the last few years have seen a dramatic increase in applied research, aimed primarily in two directions: fish culture and environmental toxicology. Physiological research is of vital importance in both these fields, and basic fish physiology is a necessary base for the applied research.

This book is intended for a wide readership among senior undergraduate, postgraduate and research students, as well as university teachers and researchers in zoology, physiology, aquaculture and biology generally. The book focuses on five major areas of basic and applied research: haemopoiesis, acid-base regulation, circulation, gastro-intestinal functions and physiological toxicology. The chapters will serve as introductions to these fields, as well as up-to-date reviews of the most recent advances in the research areas.

The idea for this book arose at the Fourth Symposium on Fish Physiology, held at Göteborg in August 1985, and most of the authors were selected from the participants in this symposium. The editors would like to thank the contributors to the book for their fine work, and for getting their manuscripts in on time.

The book is dedicated to Professor Emeritus Ragnar Fänge, who for more than 35 years has been a leading fish physiologist internationally. Of his numerous contributions in different areas of comparative physiology, his works on the functions of the teleost swimbladder and his more recent studies on haemopoiesis in fish have all rapidly become much quoted classics in fish physiology.

Stefan Nilsson
Susanne Holmgren

CONTRIBUTORS

T. Andersson, Postdoctoral Fellow, Department of
Zoophysiology, University of Göteborg, PO Box 250 59,
S-400 31 Göteborg, Sweden

P.J. Butler, Professor, Department of Zoology and Comparative
Physiology, University of Birmingham, PO Box 363,
Birmingham B15 2TT, England

G. Dave, Associate Professor, Department of Zoophysiology,
University of Göteborg, PO Box 250 59, S-400 31 Göteborg,
Sweden

P.S. Davie, Lecturer, Department of Anatomy and Physiology,
Massey University, Palmerston North, New Zealand

C. Daxboeck, Scientific and Executive Director, Pacific Gamefish
Research Foundation, PO Box 3189, Kailua-Kona, Hawaii
96745, USA

R. Fänge, Professor Emeritus, Department of Zoophysiology,
University of Göteborg, PO Box 250 59, S-400 31 Göteborg,
Sweden

L. Förlin, Associate Professor, Department of Zoophysiology,
University of Göteborg, PO Box 250 59, S-400 31 Göteborg,
Sweden

H. Gesser, Associate Professor, Department of Zoophysiology,
University of Aarhus, DK-8000 Aarhus C, Denmark

D.J. Grove, Senior Lecturer, Marine Science Laboratories,
University College of North Wales, Menai Bridge, Gwynedd
LL59 5EH, Wales

C. Haux, Postdoctoral Fellow, Department of Zoophysiology,
University of Göteborg, PO Box 250 59, S-400 31 Göteborg,
Sweden

N. Heisler, Professor, Abteilung Physiologie, Max-Planck-Institut
für Experimentelle Medizin, Hermann-Rein-Strasse 3, D-3400
Göttingen, West Germany

S. Holmgren, Associate Professor, Department of Zoophysiology,
University of Göteborg, PO Box 250 59, S-400 31 Göteborg,
Sweden

B. Holstein, Associate Professor, Department of Zoophysiology,

University of Göteborg, PO Box 250 59, S-400 31 Göteborg, Sweden

K. Johansen, Professor, Department of Zoophysiology, University of Aarhus, DK-8000 Aarhus C, Denmark

A.-C. Jönsson, Postdoctoral Fellow, Department of Zoophysiology, University of Göteborg, PO Box 250 59, S-400 31 Göteborg, Sweden

Å. Larsson, Associate Professor, Swedish Environment Protection Board, Brackish Water Toxicology Laboratory, Studsvik, Nyköping, Sweden

S. Nilsson, Professor, Department of Zoophysiology, University of Göteborg, PO Box 250 59, S-400 31 Göteborg, Sweden

P.-E. Olsson, Research Student, Department of Zoophysiology, University of Göteborg, PO Box 250 59, S-400 31 Göteborg, Sweden

J.A. Riegel, Reader, School of Biological Sciences, Queen Mary College, University of London, Mile End Road, London E1 4NS, England

1 PHYSIOLOGY OF HAEMOPOIESIS

Ragnar Fänge

Production of new blood cells is called haemopoiesis. From a general point of view this term could also be taken to include the formation of the complex solution of inorganic and organic substances, the plasma, in which the blood cells are suspended. The blood plasma continuously exchanges substances with the tissues. The liver manufactures most of the plasma proteins, but immunoglobulins are produced by plasma cells, widely distributed in the organism, and minor protein components of the plasma such as certain enzymes may be produced by extrahepatic cells or tissues. Lysozyme, a glycosidase with antibacterial activity, is produced by granulocytes and monocytes. Antifreeze glycoproteins, occurring in the plasma of Antarctic fishes at concentrations as high as 3 per cent, are synthesised in the liver (Hudson, De Vries and Haschemeyer 1979). The main plasma proteins of many fishes are albumin and globulins, which are probably produced by the liver, but several fishes are claimed to lack albumin (Sulya, Box and Gunter 1961). One could expect very low colloid osmotic pressures in the blood of those analbuminaemic fishes.

The total amount of leucocytes in the blood of fishes does not vary much, but the concentration of red cells, evaluated as the haematocrit, ranges from close to zero (Antarctic icefish, *Chaenocephalus*; leptocephalus larvae) to well above 50 per cent in active fast-swimming forms.

The main lines of blood cells in fishes are erythrocytes, granulocytes and lymphocytes. Spindle cells, or thrombocytes, may have relationships to lymphocytes. Monocytes, relatively rare blood cells, show relations to both lymphocytes and granulocytes, and to tissue macrophages.

Fishes, like mammals, possess subclasses of lymphocytes which may be demonstrated by use of mitogens (lectins). However, it is not clear whether subclasses of fish lymphocytes are fully comparable to B and T cells of mammals (McCumber, Sigel, Trauger and Cuchens 1982). Fish blood often contains blast cells, immature blood cells with a basophilic cytoplasm. The classifi-

1

cation of fish white blood cells is full of unsolved problems, partly because methods used in mammalian haematology often do not give good results when practised on fish blood (Ellis 1977).

In a living fish the concentration of blood cells is relatively constant. This cell homeostasis is due to a dynamic equilibrium between new formation (haemopoiesis) and destruction or elimination of blood cells.

Stem Cells

Vertebrate blood cells originate from multipotent or uncommitted stem cells, which are localised in haemopoietic tissues and also circulate in the blood. The stem cells have the ability to reproduce. The identity of the stem cells is the subject of much discussion. Morphologically they are similar to small and medium lymphocytes, and according to Jordan and Speidel (1924) small lymphocytes of fishes, amphibians and reptiles may function as progenitors, or stem cells, for both erythrocytes and leucocytes. Recent studies show stem cells to be identical with transitional cells, which are lymphocyte-like cells with varying degrees of basophilia and a leptochromatic nucleus (Yoffey 1985). Stem cells are transported by the circulation to different parts of the body, where they may settle at specific sites in the connective tissue adjacent to blood vessels.

Multipotent stem cells, when influenced by appropriate stimuli, differentiate into unipotent (committed) stem cells. These may multiply by repeated mitoses, undergo further differentiation and give rise to clones of blood cells: erythrocytes and various types of leucocytes.

The modern concept of stem cells is mainly founded on observations on cells from non-irradiated donor animals which, when injected into irradiated animals, form colonies or nodules in the spleen (Weiss 1981). Although those experiments were performed on mammals (rodents), the results are probably applicable in principle even to fishes. However, it is highly desirable that analogous experiments be repeated on fishes.

Maturation Stages of Blood Cells

The development of erythrocytes during teleostean haemopoiesis has been described by Iorio (1969), Lehmann and Stürenberg (1975), Härdig (1978) and others. Five or six successive steps of developing erythrocytes are distinguished. The earliest identifiable cells of the erythrocyte line, the proerythroblasts, are round with an intensely basophilic cytoplasm which stains blue with May-Grünwald-Giemsa and similar techniques, and a large, central, circular nucleus. The proerythroblasts progressively lose RNA from the cytoplasm, and synthesise haemoglobin; the nucleus gets more dense and the cell outline changes from circular to elliptic. After having passed through the sequence (1) proerythroblast, (2) young erythroblast, (3) old erythroblast, (4) polychromatic erythroblast (reticulocyte, proerythrocyte), the cells transform into (5) young mature erythrocytes and finally (6) old mature erythrocytes. In teleosts the erythrocytes are delivered from the haemopoietic tissues at the stages (3), (4) or (5). The final maturation, including synthesis of haemoglobin, takes place to a large extent in the circulating blood. The synthesis of haemoglobin has been followed by the incorporation of radioactive iron (Hevesy, Lockner and Sletten 1964; Härdig 1978). Proerythrocytes contain remnants of RNA, which may be demonstrated by supravital staining with basic dyes such as cresyl blue. Red cells showing RNA residuals are found in the blood of many fishes (Dawson 1933). In cyclostomes and elasmobranchs the release of erythrocytes from the haemopoietic sites may take place at earlier stages of maturation than in teleosts.

The maturation of granulocytes proceeds through successive more or less well defined haematological stages similar to those of erythrocytes. The early granuloblasts and lymphoblasts are difficult to distinguish from early erythroblasts. The morphology is almost the same in early stages of blood cell development.

Haemopoietic Tissues

Blood cells are formed within lymphomyeloid tissues (alternative terms: haemopoietic, lymphohaemopoietic, lymphoid, lymphatic, reticuloendothelial, etc. tissues). These are dynamic structures, which may change in size and shape, during certain conditions almost disappearing and then reappearing, in living animals.

Probably for phylogenetic reasons lymphomyeloid tissues occur quite regularly at special anatomical sites, which may differ between representatives of separate vertebrate groups. The lymphomyeloid tissues, although widely scattered in the organism, are connected into a unit, the 'lymphomyeloid complex', by migrating streams of lymphocytes and stem cells (Yoffey 1985). In mammals the lymphomyeloid tissues constitute about 3-3.5 per cent or more of the body weight (Yoffey and Courtice 1970). In the human body the bone marrow alone might weigh 1500-4000 g, or about as much as the liver (Weiss 1972). In fishes the total amount of lymphomyeloid tissues is about 0.5-1.7 per cent of the body weight (Fänge 1977). The relatively small amounts of these tissues in fishes may be seen in relationship to a relatively slow turnover of red cells, and less advanced immune functions. Fishes lack bone marrow and lymph nodes, and the cyclostomes moreover lack thymus and a true spleen, but necessary amounts of lymphomyeloid tissues nevertheless are present in all fishes.

In addition to being active in haemopoiesis, lymphomyeloid tissues function in the storage of blood cells and in destruction of effete blood cells, and play an important role in immune reactions.

Haemopoiesis in the Blood

Lymphomyeloid tissues in hagfishes are situated in the submucosa of the intestine and in the pronephros. The intestine is mainly granulopoietic, and the pronephric lymphomyeloid cell mass is a complex mixture of cells (Mattisson and Fänge 1977; Zapata, Fänge, Mattisson and Villena 1984). However, in *Myxine* the main part of the erythropoiesis appears to take place in the circulating blood, which constitutes almost a fluid bone marrow. The blood contains erythroblasts in all stages of development, often with mitotic figures. A great deal of granuloblasts and precursors of spindle cells also occur in the blood. The earliest stem cells of the erythroid line may emanate from lymphomyeloid tissues of the intestine or the pronephros, but later stages of developing red cells occur in the blood.

Erythropoietic Tissues

The spleen is the main erythropoietic tissue in elasmobranchs (sharks, rays), holocephalans (rabbit fish: *Chimaera*) and a few teleosts (*Perca*: Catton 1951; *Scorpaena*: Fey 1965). In most teleosts (bony fishes), in chondrosteans (sturgeons, paddle fish) and in holosteans (gars, bow fin) erythrocytes are produced within the kidney. In teleosts often the anterior part of the kidney (head kidney or 'pronephros') is transformed into a complex lymphomyeloid tissue, which also contains chromaffin and adrenocortical endocrine elements. In some forms (salmonids) haemopoietic tissue is spread between the tubules of the whole kidney (Catton 1951; Lehmann and Stürenberg 1975).

Granulopoietic Tissues

Granulocytes constitute about half of the leucocytes of the blood of fishes, but still larger numbers of granulocytes may be found in tissues such as the intestine and various lymphomyeloid tissues. The granulocytes are very important cells in the defence against microbes and parasites. They have a short lifespan. In mammals they occur as free cells in the circulation only for a few hours. In tissues they may live for one or two weeks. Effete granulocytes are rapidly eliminated by macrophages. The turnover rate of granulocytes is very large, and granulopoiesis has been characterised as 'one of the most dynamic activities of the organism'. In the adult mammal the daily production of granulocytes is about 1.6×10^9 granulocytes per kilogram body weight (Perry 1971). The human bone marrow contains three or four times more granulocytes than erythroblasts, which perhaps illustrates the great intensity of granulocyte production (Weiss 1972).

In the teleosts (bony fishes) granulocytes originate together with erythrocytes in renal lymphomyeloid tissue, but in other fish groups granulocytes and erythrocytes may develop in separate tissues. In elasmobranchs (sharks, rays) tissues producing mainly granulocytes and lymphocytes are found in the oesophagus (Leydig organ) and associated with the gonads (epigonal organ) (Figures 1.1, 1.3 and 1.4). These leucocyte-producing tissues may reach a very large size. Thus in a Greenland shark (*Somniosus microcephalus*) with a body weight of about 200 kg, the Leydig organ weighed more than 1 kg (Fänge and Bergström 1981), and in other large sharks the epigonal organ may weigh several kilograms (Fänge 1977, 1984).

Predominantly granulopoietic tissue occupies cavities in the cranium and the shoulder girdle of the rabbit fish (*Chimaera, Hydrolagus*) (Figure 1.5). In chondrosteans (sturgeons, paddle fish) and holosteans (bow fin, gars) bone-marrow-like granulopoietic tissue is localised in the meninges above the hind end of the

Figure 1.1: Dissection of Dogfish, *Scyliorhinus canicula* (Male, Body Weight 840 g) Showing the Epigonal Organ (EO) and the Position of the Leydig Organ within the Oesophagus (Arrows). L, liver; PF, pelvic fin; R, rectal gland; S, stomach. Bar = 1 cm

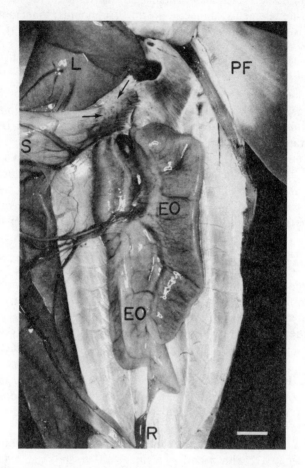

brain (Scharrer 1944; Fänge 1984). The presence of considerable amounts of lymphomyeloid tissue at this site may be phylogenetically old. The presence of a characteristic depression or channel in the hind part of the cranium indicates that certain extinct Devonian fishes had meningeal lymphomyeloid tissue (Bjerring 1984).

Extracts of the granulopoietic tissue of elasmobranchs show high activities of hydrolytic enzymes such as glycosidases. The

Figure 1.2: Heart of White Sturgeon, *Acipenser transmontanus* (Body Weight 33 kg), Left Side View. The ventricle and the truncus arteriosus (T) are covered by nodular lymphoid tissue. In the middle of the figure are seen nodules containing fat tissue (white). A, Atrium. Bar = 1 cm

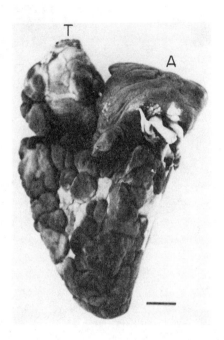

enzymes may emanate from cytoplasmic granules of the leucocytes (Fänge, Lundblad, Slettengren and Lind 1980). It is not known whether there are any mechanisms allowing leucocytic enzymes to be released from lymphomyeloid tissues.

Pericardial Tissue of Sturgeons

A prominent whitish or reddish nodular tissue is found in the pericardium of the heart of sturgeons (Figure 1.2) and paddlefish (*Polyodon*). A thorough histological work was published by Hertwig (1873). This tissue, noticed by anatomists 300 years ago, is sometimes called the pericardial organ. It resembles structurally quite closely the avian lymph nodes (Drzewina 1905). Hertwig (1873) and Scatizzi (1933) suggested that within the pericardial tissue lymphocytes originate by transformation of vascular endothelial cells. According to my own observations (Fänge 1986), lymphocytes are migrating through the endothelium of

Figure 1.3: *Etmopterus spinax* (Deep-water Shark), Section through the Oesophagus. The Leydig organ forms two large white portions. These are situated between the muscularis (white in the figure) and the mucosa (dark in the figure). Bar = 1 cm. Reproduced with permission from Fänge (1968)

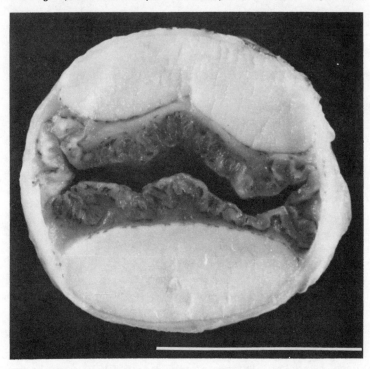

venous sinuses associated with lymphoid cell masses (Figure 1.6). The endothelial cells have a light, vacuolated cytoplasm and some kinds of interaction may take place between the migrating lympho-cytes and the endothelial cells (Figure 1.7). The lymphoid cell masses of the pericardial tissue have a complex and confusing structure. Lymphocytes may be seen inside other cells (empero-polesis). The interactions between lymphocytes and endothelial cells may be functionally important. In mammalian lymph nodes endothelial cells, stimulated by migrating lymphocytes, appear to secrete substances which provoke accumulation of lymphocytes (Baldwin 1981).

The biological significance of the unique lymphoid tissue of the pericardium of sturgeons is unknown, but the tissue probably functions in immune processes and perhaps in lymphopoiesis.

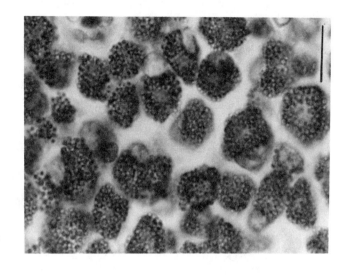

Figure 1.5: Pacific Rabbit Fish, *Hydrolagus colliei.* Histological section through the meningeal myeloid tissue showing eosinophilic granulocytes. Giemsa. Bar = 10μm

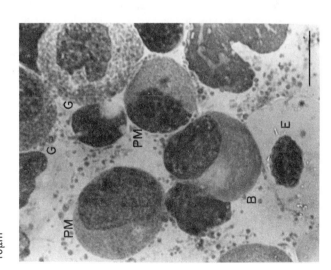

Figure 1.4: Dogfish, *Scyliorhinus canicula.* Imprint of the epigonal organ, stained with May-Grünwald-Giemsa. E, Erythrocyte; G, granulocytes, probably eosinophils; PM, promyelocytes; B, blast cell (or plasma cell). Bar = 10μm

Figure 1.6: White Sturgeon, *Acipenser transmontanus*. Histological section through a small lobe of the pericardial lymphoid tissue. The lymphoid cell mass is surrounded by a venous sinus (VS) bordered by endothelium (not clearly seen at the low magnification). HW, Heart wall. Eosin-haematoxylin. Bar = 100 μm

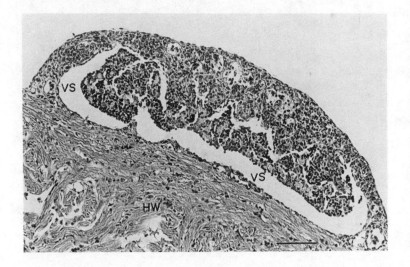

Macrophages, Antigen-trapping Cells, Plasma Cells

Lymphomyeloid tissues contain numerous cells participating in immune responses. Macrophages are important in the first contact with antigen-carrying foreign cells or particles. The anterior part of the kidney has been used as material for the separation and characterisation of macrophages from salmonid fishes (Braun-Nesje, Bertheussen, Kaplan and Seljlid 1981). The function of the so-called melanomacrophages is insufficiently understood. Melanomacrophage centres in lymphomyeloid tissues of teleosts have been compared to germinal centres of lymphoid tissues of higher vertebrates, but this analogy is not yet fully verified (Agius 1985). Dendritic macrophages and cells of the ellipsoid sheaths of the spleen may play a role in handling antigenic material (Ellis 1980; White 1981; Fulop and McMillan 1984). Plasma cells, which are producing antibodies (immunoglobulins), are found in lymphomyeloid tissues. In teleosts plasma cells are abundant in the head kidney (Smith, Wivel and Potter 1970). In elasmobranchs the spleen harbours plasma cells; these also occur in small numbers in the Leydig and epigonal organs (Fänge and Mattisson 1981; Fänge

Figure 1.7: White Sturgeon, *Acipenser transmontanus*. Histological section through the endothelium covering the lymphoid mass. Higher magnification than in Figure 1.6. Lymphocytes are seen in the lumen of the venous sinus and migrating through or infiltrating the endothelial layer. EN, Endothelial nuclei; C, light vacuolated cytoplasm of endothelial cells. The endothelial surface is enlarged by irregular projections of endothelial cell cords. Eosin-haematoxylin. Bar = 10 μm

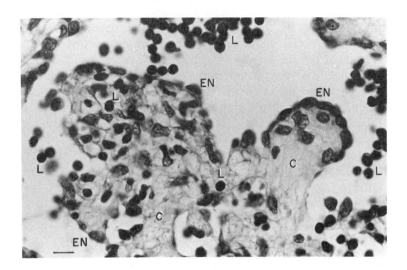

and Pulsford 1983). Hagfish (*Myxine*), which has poorly developed immune responses, contains scattered plasma cells in the lymphoid mass of the pronephros (Zapata *et al.* 1984).

Synthesis of Haemoglobin

Haemoglobin is formed inside maturing red cells by the combination of an iron-porphyrin compound, haeme, with a peptide, globin.

The precursor of haeme, δ-aminolevulinic acid (ALA) is synthesised within the mitochondria of erythroblasts by condensation of glycine and succinic acid (rather succinyl coenzyme A) under the influence of the enzyme δ-aminolevulinic acid synthetase (ALA-synthetase). Another enzyme, ALA-dehydratase, combines two molecules of ALA to form protoporphyrin, which together with iron (Fe^{++}) gives haeme. Globin is synthesised by cytoplasmic ribosomes.

The iron originally comes from the food. Because the solubility of inorganic iron increases at low pH, gastric secretion of acid probably facilitates the resorption of iron (Granick 1953). A few fishes lack stomach and production of hydrochloric acid (rabbit-fish, *Chimaera*, is one example). It is not known how iron resorption in these fishes is affected by the stomach-less condition. Iron resorbed by the intestine is stored as ferritin, an iron-containing protein. In the dogfish (*Scyliorhinus canicula*) dark-red crystals of ferritin have been isolated from the intestine and the liver, but none was found in the spleen (Tecce 1952). Iron is transported in the organism bound to a plasma globulin, trans-ferrin. There are several transferrins, which constitute a family of related proteins. Even the primitive vertebrate *Myxine* contains transferrin in its blood plasma (Palmour and Sutton 1971). Erythroblasts possess on their surface specific receptors for trans-ferrin, which enables the cells to take up iron (Fletcher and Huehns 1968). Uptake of iron by erythroblasts takes place both within erythropoietic tissues and in the circulating blood. Experiments using radioactive iron show that in fishes a considerable uptake of iron may occur in the blood (*Tinca tinca*: Hevesy *et al.* 1964; *Lepisosteus*: McLeod, Sigel and Yunis 1978). This is in accordance with morphological results indicating that immature erythrocytes are common in the blood of fishes, especially cyclostomes and elasmobranchs but also many teleosts (Dawson 1933; Kanesada 1956; Mattisson and Fänge 1977).

Erythrocytes are highly specialised cells with a limited lifespan. Human red cells survive in the blood for about 120 days. Red cells of non-mammalian vertebrates including fishes are nucleated, less specialised, and presumably have longer survival times than mammalian red cells. Hevesy *et al.* (1964) found a lifespan of more than 150 days for erythrocytes of a teleost (*Tinca tinca*).

Aged red cells get fragile and are withdrawn from the cir-culation by macrophages situated in lymphomyeloid tissues, especially the spleen. During phagocytosis the haeme of haemo-globin is split into porphyrin and free iron. The porphyrin is con-verted into bile pigments and excreted by the liver, and the iron is converted into storage form (ferritin or haemosiderin) or transport form (transferrin), and most of it is re-utilised for haemoglobin synthesis. So called melanomacrophage centres of lymphomyeloid tissues of teleosts contain haemosiderin, probably derived from red cells destroyed by macrophage activity (Agius 1985). The iron metabolism of the vertebrate organism forms an almost closed system. Only small amounts of iron pass in and out through the intestinal mucosa.

The amount of iron being transported in the plasma is small compared with that bound in haemoglobin, or in storage protein of the spleen and other tissues (Table 1.1).

Microcirculation and Microenvironments

Factors such as pH, hormones, carbon dioxide and oxygen tensions may influence growth and differentiation of haemopoietic cells. The microcirculation is of great importance for the micro-environment. Jordan (1933) concluded from comparative studies of blood formation in different vertebrates that erythropoiesis is associated with a profuse but sluggish, relatively stagnant, sinusoidal venous circulation. He thought high carbon dioxide tension to be of prime importance. Granulocytes, according to the same author, differentiate within sparsely vascularised mesenchymal connective tissue.

By a special technique Brånemark (1959) was able to observe under the microscope the circulation in living haemopoietic bone marrow of a mammal (rabbit). The blood circulates through a system of venous sinuses, which vary rhythmically in degree of dilation. *In vivo* observations of the blood circulation in haemopoietic tissues of fishes have not been made. All available information comes from ordinary anatomical and histological studies. The erythrocyte-producing renal tissue of teleosts and chondrosteans is supplied with blood from renal portal veins (teleosts:

Table 1.1: Concentration of Iron in the Blood and Tissues of the Tench (*Tinca tinca*) in μg Fe per 100 g Wet Weight. (Calculated from van Dijk, Lagerwerf, van Eijk and Leijsne 1975.)

Tissue	Concentration of iron
Plasma	16
Erythrocytes	1170
Spleen	12900
Liver	710
Kidney	180

Sharma 1969; *Polyodon*: Downey 1909). Catton (1951) observed that in the teleostean renal intertubular haemopoietic tissue, endothelium-lined venous channels communicate with capillaries. The endothelium seemed to be fenestrated allowing blood cells to pass through the walls of the channels. The blood supply from renal portal veins through sinuses or channels may provide a relatively sluggish circulation, which would favour erythropoiesis in the sense of Jordan (1933). Granulopoietic tissues like the Leydig and epigonal organs of elasmobranchs are richly supplied with venous sinuses but have a poor afferent arterial blood supply (Figures 1.8A, B). The same is true of the lymphomyeloid tissues of rabbit fishes (*Chimaera, Hydrolagus*).

Accumulation and multiplication of lymphocytes may be associated with filtration of body fluids through membranes. Hertwig (1873) noticed that lymph nodes as a rule take their origin in the adventitia of blood vessels. Accumulation of lymphocytes occurs in the perivascular sheaths of blood vessels in the spleen. In the spleen of teleosts periarterial lymphoid sheaths are poorly developed (Fänge and Nilsson 1985), but prominent lymphoid sheaths (white pulp) are found in the spleen of sturgeons (Figure 1.8D; Fänge 1986). A primitive splenic vascular architecture is found in *Erpetoichthys* (rope fish: Brachiopterygii). In this fish a splenic artery and vein are surrounded by lymphoid tissue (white pulp), outside which is a layer of red pulp (Yoffey 1929). However, this simplified vascular organisation of the spleen may be an example of anatomical reduction associated with the snake-like body shape of *Erpetoichthys*. It would be interesting to investigate the structure of *Polypterus* (bichir), a less specialised form of brachiopterygian fish.

The lymphomyeloid tissue of the intestine of hagfishes (*Myxine, Eptatretus*) is often considered a primitive spleen (Tomonaga, Hirokane, Shinohara and Awaya 1973). This tissue consists of haemopoietic cell islands grouped in a star-like manner around irregular branches of intestinal portal veins. The tissue is granulopoietic rather than lymphopoietic (Figure 1.8C).

Large numbers of lymphocytes and other leucocytes are regularly found in the intestinal mucosa. Resorption of water and substances by the mucosal epithelium may sustain fluid transport favouring accumulation of lymphocytes. Large follicle-like lymphocytic accumulations occur in the spiral valve of the intestine of the paddle fish (*Polyodon*) (Weisel 1973) and sturgeons

Figure 1.8: Diagram Showing at Low Magnification Microcirculatory
Vessels in Histological Sections of Various Lymphomyeloid Tissues in
Fishes. A. Dogfish, *Scyliorhinus canicula*, Leydig organ. Lobes of
lymphomyeloid tissue are stretching between the mucosa and the
muscularis of the oesophagus. Surrounding the lobes are thin-walled
venous sinuses. One single transversal sectioned artery is indicated in the
middle of the figure. (From a specimen injected with India ink via the
intestinal portal vein; Fänge and Pulsford 1983.) B: Dogfish, *Scyliorhinus
canicula*, epigonal organ. Irregular thin-walled veins or sinuses are
penetrating lymphomyeloid tissue. C: Intestinal lymphomyeloid tissue
('primitive spleen') of hagfish, *Myxine glutinosa*. Granulopoietic cell islands
are associated around branches of the intestinal portal vein. No cell islands
are found around arteries. The blood vessels and the lymphomyeloid cell
islands are embedded in fat tissue. The fat cells are very large. D: White
sturgeon, *Acipenser transmontanus*, spleen. Lymphoid sheaths (white pulp)
are found around arteries and veins, but the ellipsoid arterioles are
surrounded by red pulp. Eosinophilic granulocytes are abundant at the
periphery of white pulp (indicated by dots). a, Artery; v, vein; dotted areas:
lymphomyeloid tissue; hatched areas: red pulp

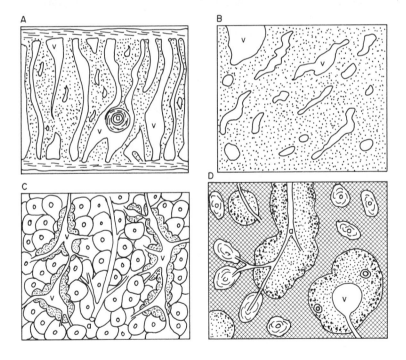

(*Acipenser*). The follicle-like lymphoid masses in the intestine of chondrosteans may have relationships to infections by microbes or parasites.

The thymus, a purely lymphoid tissue, is found in all fishes except cyclostomes. Hormones produced by reticular epithelial cells may be important for thymic functions. The blood circulation of the thymus in fishes is poorly known.

Hormonal Control of Haemopoiesis

Unipotent stem cells, committed for erythropoiesis, are assumed to have receptors for the hormone erythropoietin. This hormone stimulates the cells to synthesise haemoglobin. Erythropoietin has been demonstrated in both mammals and non-mammalian vertebrates. Several observations indicate that in fishes erythropoiesis is stimulated by humoral factors, probably erythropoietin. Experimental anaemia caused by bleeding or by injection of phenylhydrazine stimulate erythropoiesis in fishes (Cameron and Wohlschlag 1969; Smith, McLain and Zaugg 1971). In the tropical fish, gourami (*Trichogaster trichopterus*), blood plasma from experimentally anaemic fish stimulates erythropoiesis and synthesis of haemoglobin when injected into non-anaemic fish. Mammalian urinary erythropoietin also stimulates erythropoiesis in this fish but only in high doses (Zanjani, Yu, Perlmutter and Gordon 1969; Yu, Kiley, Sarot and Perlmutter 1971). McLeod *et al.* (1978) observed that in the gar (*Lepisosteus platyrhincus*) erythropoiesis is stimulated by experimental anaemia but not by hypoxia. Apparently in this species hypoxia alone does not induce production of erythropoietin. Anaemia produced by bleeding or otherwise stimulates formation or release of erythropoietin in most vertebrates studied.

In addition to erythropoietin certain other humoral factors may exist, which control haemopoiesis. The existence of a leucocytopoietin or granulopoietin has been postulated. Mature blood cells, such as full-grown granulocytes, may produce substances, chalones, which inhibit production of new cells. Thus hypothetical granulocytic chalone may have a negative feedback effect on granulopoiesis (Perry 1971). On the other hand, aged granulocytes are supposed to liberate substances which promote the formation of new granulocytes (Teir 1966).

Various Effects Influencing Haemopoiesis in Fish

Many lymphomyeloid structures are very sensitive to adreno-cortical hormones. The spleen of the teleost *Astyanax* was depleted of lymphocytes when the fishes were subjected to stress. Administration of adrenocorticotropic hormone (ACTH) decreased the number of lymphocytes in the head kidney, and oedematous spaces filled with acellular material appeared. The changes were reversible within 1-2 days (Rasquin 1951). On the other hand Pandey, Chanchal and Singh (1978) observed that treatment of the air-breathing fish *Anabas testudineus* with hydrocortisone, testosterone, adrenalin or noradrenalin caused an increase in the number of circulating red blood cells. But the mechanism of the changes in blood parameters was not found. Release of stored erythrocytes from the spleen might have contributed to the effects.

The enzyme delta-aminolevulinic acid dehydratase (ALA-D), which catalyses the formation of porphobilinogen during haemo-globin synthesis, is very sensitive to lead. Johansson-Sjöbeck and Larsson (1979) showed that exposure of rainbow trout (*Salmo gairdneri*) to sublethal doses of inorganic lead strongly depressed the activity of ALA-D in blood and haemopoietic tissues. High lead concentrations caused anaemia. The inhibition of ALA-D in fish may be a sensitive test to detect lead exposure in nature.

Exposure of flounder (*Pleuronectes flesus*) to sublethal doses of inorganic cadmium caused reduced haematocrit, haemoglobin level and red blood cell count. The anaemia was accompanied by a significant increase of the ALA-D activity in renal tissue (Johansson-Sjöbeck and Larsson 1978; Houston and Keen 1984).

Haemopoietic tissues are sensitive to irradiation, but in fishes the erythropoiesis is considerably more radio-resistant than in mammals (Sletten, Lockner and Hevesey 1964). This may in part be a temperature effect. At the low body temperature of fishes compared with that of mammals, lesions produced by irradiation are less pronounced than at higher temperatures. In hagfish (*Myxine*), irradiation caused a decrease of leucocytes in the blood. An unusual resistance towards irradiation was observed in this animal (Finstad, Fänge and Good 1969).

Summary

Haemopoiesis occurs in many different tissues of fishes (Table 1.2). Suitable microenvironments for the multiplication and differentiation of stem cells, and for the accumulation of leucocytes, may be realised at different anatomical sites. Studies on haemopoiesis in fishes may add data useful in understanding cellular differentiation processes, interactions between lymphocytes and endothelial cells, and immune mechanisms.

Follicle-like accumulations of lymphocytes, although less prominent than in mammals and birds, are observed in fishes. Especially distinct lymphocytic accumulations are found in the intestine and the spleen of chondrosteans (sturgeons, paddle fish).

The role of the vascular architecture for the microenvironments of haemopoietic tissues is worth investigating. In spite of ingenious hypotheses (Jordan 1933), it is not clear why certain types of blood cells aggregate and reproduce at specific sites. The anatomical separation of erythropoiesis and granulopoiesis in elasmobranch, holocephalan and ganoid fishes invites biochemical studies of mature and immature erythrocytes and leucocytes. Enormous amounts of granulocytes and their precursors are found in the Leydig and epigonal tissues of elasmobranchs. These tissues are very rich in hydrolytic enzymes, probably contained in leucocytic lysosomes (Fänge *et al.* 1980). The mechanisms for the release of glycolytic enzymes from leucocytes ought to be examined. If released within the organism, glycosidases may be thought to change the chemical structures of cell-surface glycoconjugates. Those substances are assumed to play a role in directing the traffic of migrating cells (Hooghe and Pink 1985).

Phylogenetic approaches may be rewarding. The organisation of the spleen in sturgeons is more complex than in most other fishes, and approaches conditions in higher vertebrates. Further studies of the structure and function of the fish spleen (Fänge and Nilsson 1985) and other haemopoietic tissues in fishes may offer much of interest. Special problems in spleen physiology, for instance the function of the ellipsoidal arterioles, may be investigated with advantage in certain types of fishes.

Table 1.2: Distribution in the Body of Lymphomyeloid Tissues in Fishes of Different Groups. (For references see Fänge 1982, 1984; from various sources)

Group	Thymus	Spleen	Intestine	Oesophagus	Kidney	Meninges/ neural arch	Skeleton	Heart
Myxini (hagfish)	0	(+)	+	0	+	0	0	0
Petromyzontidae (lampreys)	0	(+)	+	0	+	+	(+)	0
Holocephali (rabbit fish)	+	+	+	0	0	0	+	0
Elasmobranchii (sharks, rays)	+	+	+	+/0	0(+)	0(+)	0	0
Coelacanth	+	+	+	0	+	0	0	0
Dipneusti (lungfish)	+	+	+	0	+	0	0	0
Brachiopterygii (bichirs)	+	+	+	0	?	(+)	0	0
Chondrostei (sturgeons)	+	+	+	0	+	+	0	0
Holostei (gars, bow fin)	+	+	+	0	+	+	0	+
Teleostei (bony fish)	+	+	+	(+)/0	+	0	0	0

+ = Present; (+) = weakly represented or doubtful; 0 = not present. Lymphomyeloid tissue of the intestine of cyclostomes is often regarded as representing a primitive spleen. Note: small amounts of lymphomyeloid elements may occur in the liver, pancreas and other tissues of fishes (Drzewina 1905); knowledge about many fish groups is incomplete, and the table may need to be revised in the future.

References

Agius, C. (1985) 'The Melano-macrophage Centres of Fish: a Review' in M.J. Manning and M.F. Tatner (eds), *Fish Immunology*, Academic Press, New York

Baldwin, III, W.M. (1981) 'The Symbiosis of Immunocompetent and Endothelial Cells', *Immunology Today*, *3*, 267-8

Bjerring, H.H. (1984) 'The Term "Fosse Bridgei" and Five Endocranial Fossae in Teleostome Fishes', *Zoologica Scripta*, *13*, 231-8

Braun-Nesje, R., Bertheussen, K., Kaplan G. and Seljlid, R. (1981) 'Salmonid Macrophages: Separation, *in vitro* Culture and Characterization', *Journal of Fish Disease*, *4*, 141-51

Brånemark, P. (1959) 'Vital Microscopy of Bone Marrow in Rabbit', *Scandinavian Journal of Clinical and Laboratory Investigations*, *11* (Supplement 38), 1-82

Cameron, J.N. and Wohlschlag, D.E. (1969) 'Respiratory Response to Experimentally Induced Anemia in the Pinfish (*Lagadon rhomboides*)', *Journal of Experimental Biology*, *50*, 307-17

Catton, W.T. (1951) 'Blood Cell Formation in Certain Teleost Fishes', *Blood*, *6*, 39-60

Dawson, A.B. (1933) 'The Relative Number of Immature Erythrocytes in the Circulating Blood of Several Species of Marine Fishes', *Biological Bulletin*, *64*, 33-43

Dijk, J.P. van, Lagerwerf, A.J., Eijk, H.G. van and Leijsne, B. (1975) 'Iron Metabolism in the Tench (*Tinca tinca* L.) I. Studies by Means of Intravascular Administration of ^{59}Fe (III) Bound to Plasma', *Journal of Comparative Physiology*, *99*, 321-30

Downey, H. (1909) 'The Lymphatic System of the Kidney of *Polyodon spathula*', *Folia Haematologica*, *8*, 415-66

Drzewina, A. (1905) 'A l'étude du tissu lymphoïde des Ichtyopsidés', *Archives de Zoologie Expérimentale et Génerale*, *33*, 145-338

Ellis, A.E. (1977) 'The Leucocytes of Fish: a Review', *Journal of Fish Biology*, *11*, 453-91

Ellis, A.E. (1980) 'Antigen-trapping in the Spleen and Kidney of the Plaice *Pleuronectes platessa* L.', *Journal of Fish Disease*, *3*, 413-26

Fänge, R. (1968) 'The Formation of Eosinophilic Granulocytes in the Oesophageal Lymphomyeloid Tissue of the Elasmobranchs', *Acta Zoologica (Stockholm)*, *49*, 155-61

Fänge, R. (1977) 'Size Relations of Lymphomyeloid Organs in some Cartilaginous Fish', *Acta Zoologica (Stockholm)*, *58*, 143-62

Fänge, R. (1982) 'A Comparative Study of Lymphomyeloid Tissue in Fish', *Developmental and Comparative Immunology*, Suppl. *2*, 23-33

Fänge, R. (1984) 'Lymphomyeloid Tissues in Fishes', *Videnskabelige Meddelelser fra dansk naturhistorisk Forening*, *145*, 143-62

Fänge, R. (1986) 'Lymphoid Organs in Sturgeons (Acipenseridae)', *Veterinary Immunology and Immunopathology*, in press

Fänge, R. and Bergström, B. (1981) 'Håkäring Somniosus microcephalus undersökt på Kristineberg', *Fauna och Flora*, *76*, 31-6 (Swedish; English summary)

Fänge, R. and Mattisson, A. (1981) 'The Lymphomyeloid (Hemopoietic) System of the Atlantic Nurse Shark, *Ginglymostoma cirratum*', *Biological Bulletin*, *160*, 240-9

Fänge, R. and Nilsson, S. (1985) 'The Fish Spleen: Structure and Function', *Experientia*, *41*, 152-8

Fänge, R. and Pulsford, A. (1983) 'Structural Studies on Lymphomyeloid Tissues of the Dogfish, *Scyliorhinus canicula* L.', *Cell and Tissue Research, 230,* 337-51

Fänge, R., Lundblad, G., Slettengren, K. and Lind, J. (1980) 'Glycosidases in Lymphomyeloid (Hematopoietic) Tissues of Elasmobranch Fish', *Comparative and Biochemical Physiology, 67B,* 527-32

Fey, F. (1965) 'Hämatologische Untersuchungen der blutbildenden Gewebe niederer Wirbeltiere', *Folia Haematologica, 84,* 122-46

Finstad, J., Fänge, R. and Good, R.A. (1969) 'The Development of Lymphoid Systems: Immune Response and Radiation Sensitivity in Lower Vertebrates', in *Lymphatic Tissue and Germinal Centers in Immune Response,* Plenum, New York, pp. 21-31

Fletcher, J. and Huehns, E.R. (1968) 'Function of transferrin', *Nature, 218,* 1211-14

Fulop, G.M.I. and McMillan, D.B. (1984) 'Phagocytosis in the Spleen of the Sunfish *Lepomis spp.*', *Journal of Morphology, 179,* 175-95

Granick, S. (1953) 'Inventions in Iron Metabolism', *American Naturalist, 87,* 65-75

Härdig, J. (1978) 'Maturation of Circulating Red Blood Cells in Young Baltic Salmon (*Salmo salar* L.)', *Acta Physiologica Scandinavica, 102,* 290-300

Hertwig, R. (1873) 'Die lymphatische Drüsen auf der Oberfläche des Störherzens', *Archiv für mikroskopische Anatomie, 9:* 62-79

Hevesy, G., Lockner, D. and Sletten, K. (1964) 'Iron Metabolism and Erythrocyte Formation in Fish', *Acta Physiologica Scandinavica, 60,* 256-66

Hooghe, R.J. and Pink, J.R.L. (1985) 'The Role of Carbohydrates in Lymphoid Cell Traffic', *Immunology Today, 6,* 180-1

Houston, A.H. and Keen, J.E. (1984) 'Cadmium Inhibition of Erythropoiesis in Goldfish, *Carassius auratus*', *Canadian Journal of Fisheries and Aquatic Sciences, 41,* 1829-34

Hudson, A.P., De Vries, A.L. and Haschemeyer, A.E.V. (1979) 'Antifreeze Glycoprotein Biosynthesis in Antarctic Fishes', *Comparative Biochemistry and Physiology, 62B,* 179-83

Iorio, R.J. (1969) 'Some Morphologic and Kinetic Studies of the Developing Erythroid Cells of the Common Goldfish, *Carassius auratus*', *Cell and Tissue Kinetics, 2,* 319-31

Johansson-Sjöbeck, M.-L. and Larsson, Å. (1978) 'The Effect of Cadmium on the Hematology and on the Activity of δ-Aminolevulinic Acid Dehydratase (ALA-D) in Blood and Hematopoietic Tissues of the Flounder, *Pleuronectes flesus* L.', *Environmental Research, 17,* 191-204

Johansson-Sjöbeck, M.-L. and Larsson, Å. (1979) 'Effects of Inorganic Lead on Delta-aminolevulinic Acid Dehydratase Activity and Hematological Variables in the Rainbow Trout, *Salmo gairdneri*', *Archives of Environmental Contamination and Toxicology, 8,* 419-31

Jordan, H.E. (1933) 'The Evolution of Blood-forming Tissue', *Quarterly Review of Biology, 8,* 58-76

Jordan, H.E. and Speidel, C.C. (1924) 'Studies on Lymphocytes. II. The Origin, Function, and Fate of the Lymphocytes in Fishes', *Journal of Morphology, 38,* 529-50

Kanesada, A. (1956) 'A Phylogenetic Survey of Hemocytopoietic Tissues in Submammalian Vertebrates', *Bulletin of Yamaguchi Medical School, 4,* 1-36

Lehmann, J. and Stürenberg, F.-J. (1975) 'Haematologisch-serologische Substratuntersuchungen an der Regenbogenforelle (*Salmo gairdneri* Richardson)', II. *Gewässer und Abwässer,* Heft 55/56, 123 pp. H. Kaltenmeier Söhne, Krefeld-Hüls

McCumber, L.J., Sigel, M.M., Trauger, R.J. and Cuchens, M.A. (1982) 'RES

Structure and Function of the Fishes', in N. Cohen and M.M. Sigen (eds), *The Reticuloendothelial System, Vol. 4, Phylogeny and Ontogeny,* Cambridge University Press, Cambridge, pp. 393-422

McLeod, T.V., Sigel, M.M. and Yunis, A.A. (1978) 'Regulation of Erythropoiesis in Florida Gar, *Lepisosteus platyrhincus', Comparative Biochemistry and Physiology, 60A,* 145-50

Mattisson, A. and Fänge, R. (1977) 'Light- and Electronmicroscopic Observations on the Blood Cells of the Atlantic Hagfish, *Myxine glutinosa* (L.)', *Acta Zoologica (Stockholm), 58,* 205-21

Palmour, R.M. and Sutton, H.E. (1971) 'Vertebrate Transferrins. Molecular Weights, Chemical Composition, and Iron-binding Studies', *Biochemistry, 10,* 4026-32

Pandey, B.N., Chanchal, A.K. and Singh, S.B. (1978) 'Effect of Hormones and Pharmacological Drugs on the Blood of *Anabas testudineus* (Bloch)', *Folia Haematologica, 105,* 665-71

Perry, S. (1971) 'Proliferation of Myeloid Cells', *Annual Review of Medicine, 22,* 171-84

Rasquin, P. (1951) 'Effects of Carp Pituitary and Mammalian ACTH on the Endocrine and Lymphoid System of the Teleost *Astyanax mexicanus', Journal of Experimental Zoology, 117,* 317-58

Scatizzi, I. (1933) 'L'organo linfomieloide pericardico dello storione', *Archivio Zoologico Italiano, 18,* 1-26

Scharrer, E. (1944) 'The Histology of the Meningeal Myeloid Tissue in the Ganoids', *Anatomical Records, 88,* 291-310

Sharma, S. (1969) 'The Modes of Vascular Supply in the Kidneys of Teleosts', *Annals of Zoology (Agra, India), 5,* 85-109

Sletten, K., Lockner, D. and Hevesy, G. (1964) 'Radiosensitivity of Hemopoiesis in Fish. I. Studies at 18°C', *International Journal of Radiation Biology, 8,* 317-28

Smith, A.M., Wivel, N.A. and Potter, M. (1970) 'Plasmacytopoiesis in the Pronephros of the Carp (*Cyprinus carpio*)', *Anatomical Records, 167,* 351-69

Smith, C.E., McLain, L.R. and Zaugg, W.S. (1971) 'Phenylhydrazine-induced Anemia in Chinook Salmon', *Toxicology and Applied Pharmacology, 20,* 73-81

Sulya, L.L., Box, B.E. and Gunter, G. (1961) 'Plasma Proteins in the Blood of Fishes from the Gulf of Mexico', *American Journal of Physiology, 200,* 152-4

Tecce, G. (1952) 'Presence of Ferritin in a Marine Elasmobranch, *Scyllium canicula', Nature, 170,* 75

Teir, H. (1966) 'Neubildung und Abbau der Granulozyten', *Verhandlungen der deutschen Gesellschaft für Pathologie, 50,* 219-33

Tomonaga, S., Hirokane, T., Shinohara, H. and Awaya, K. (1973) 'The Primitive Spleen of the Hagfish', *Zoological Magazine (Tokyo), 82,* 215-17

Weisel, G.F. (1973) 'Anatomy and Histology of the Digestive System of the Paddlefish (*Polyodon spathula*)', *Journal of Morphology, 140,* 243-55

Weiss, L. (1972) *The Cells and Tissues of the Immune System,* Prentice-Hall Inc., Englewood Cliffs, NJ

Weiss, L. (1981) 'Haemopoiesis in Mammalian Bone Marrow', in *Microenvironments in Haemopoietic and Lymphoid Differentiation, Ciba Foundation Symposium, 84,* 5-21

White, R.G. (1981) 'Antigen Transport in the Spleen', *Immunology Today, 2,* 150-1

Yoffey, J.M. (1929) 'Contribution to the Comparative Histology of the Spleen with Reference to its Cellular Constituents. I. In Fishes', *Journal of Anatomy, 63,* 314-44

Yoffey, J.M. (1985) 'Cellular Migration Streams. The Integration of the Lymphomyeloid Complex', *Lymphology, 18*, 5-21
Yoffey, J.M. and Courtice, F.C. (1970) *Lymphatics, Lymph and the Lymphomyeloid Complex*, Academic Press, New York
Yu, M.L., Kiley, C.W., Sarot, D.A. and Perlmutter, A. (1971) 'Relation of Hemosiderin to Erythropoiesis in the Blue Gourami, *Trichogaster trichopterus*', *Journal of the Fisheries Research Board of Canada, 28*, 47-8
Zanjani, E.D., Yu, M.-L., Perlmutter, A. and Gordon, A.S. (1969) 'Humoral Factors Influencing Erythropoiesis in the Fish (Blue Gourami — *Trichogaster trichopterus*)', *Blood, 33*, 573-81
Zapata, A., Fänge, R., Mattisson, A. and Villena, A. (1984) 'Plasma Cells in Adult Atlantic Hagfish, *Myxine glutinosa*', *Cell and Tissue Research, 235*, 685-93

2 MECHANISMS AND LIMITATIONS OF FISH ACID-BASE REGULATION

Norbert Heisler

One of the most important tasks for homeostatic regulation is maintenance of a set-point pH in the body fluids of an animal. Enzyme systems catalysing metabolic energy-producing processes usually possess pH optima in their overall activity curves, so pH changes may result in reduced metabolic performance. Accordingly, any net production of acidic or alkaline metabolic end-products has to be balanced by equimolar elimination from the body fluids. Changes of metabolic production of acid-base relevant substances, or changes in the composition of the environmental water may, however, challenge the regulatory mechanisms responsible for the acid-base regulation to an extent that transient disturbances of the equilibrium cannot be avoided.

The regulatory mechanisms available for acid-base regulation are in principle the same in all classes of animals; the extent to which they are (or can be) utilised, however, is vastly different. The special situation of fishes (from an anthropocentric viewpoint) is characterised by the intimate contact with their aqueous environment, including utilisation of water as a gas exchange medium for the majority of species. This situation provides favourable conditions for ion transfer mechanisms supporting the acid-base regulation, but also includes severe restrictions in the regulation of P_{CO_2} in the body fluids, which do not apply to terrestrial animal species. This chapter will delineate features and limitations of acid-base regulatory mechanisms in fishes. Because of space limitations this description has to be rather general, and for more detailed information the reader is referred to the cited review articles published in relevant handbooks.

Mechanisms for Acid-Base Regulation

Buffering

During the time period between introduction of surplus H^+ and OH^- ions into the body fluids by either metabolic production or by

24

influence of environmental factors and their final elimination, the acid-base status is stressed to an extent that is very much dependent on the buffering properties of the body fluids, i.e. on the capability of intracorporeal substances to transfer free H^+ ions (or OH^-) into the non-dissociated state.

Biologically relevant non-bicarbonate buffers are mainly protein residues, which are characterised by pK values close to physiological pH values, and by the fact that the total concentration (sum of the acid and base forms) of the buffer is constant (cf. Heisler 1986a). In contrast, the buffering capability of the CO_2-bicarbonate buffer system is poor at physiological pH values as long as buffering takes place in a closed buffer system (sum of the acid and base forms constant), because of its low pK value of 6.0-6.3 (depending on temperature, ionic strength and other factors, cf. Heisler 1986a). The biological importance of the bicarbonate buffer system is accordingly closely related to the fact that the concentration of the acid form of the buffer system (CO_2) is adjusted to certain set-point values, and kept quite constant in the organisms by respiratory gas exchange. The buffering capability of the bicarbonate buffer system is then logarithmically related to the bicarbonate concentration (cf. Heisler 1986a).

The buffer value as an expression of the buffering capability of a buffer system is defined as the amount of H^+ ions required to change pH by one unit ($\beta = -\Delta H^+_{bound}/\Delta pH$). The non-bicarbonate buffer values of extracellular fluid compartments are generally low due to the low protein content, whereas the values in intracellular compartments are larger by factors of 20-40 (Table 2.1; for references see Heisler 1986a). In contrast, the bicarbonate buffer value is higher in the extracellular fluid than in intracellular body compartments by factors of 2-7. The overall contribution (buffer capacity, κ) of the bicarbonate buffer, however, is limited because of the much smaller volume of the extracellular in comparison to the intracellular body compartments (Table 2.1).

Buffering is a mechanism for only transient acid-base regulation. Surplus acid-base relevant ions (e.g. H^+, OH^- or HCO_3^-) are masked by buffer bases as a result of changes in pH, which are reduced as compared with an unbuffered system. Accordingly, the mechanism is incapable of restoring the steady state conditions, but can support maintenance of pH values still compatible with life functions until the cause of the disturbance is eliminated, or until pH is compensated by other means.

Table 2.1: Buffering Properties of Various Body Compartments of Fishes and Mammals Expressed by Non-bicarbonate (β_{NB}) or Bicarbonate Buffer Values (β_{Bic}, values for steady state pH and constant P_{CO_2}). The buffer capacity (κ) of body compartments is the product of buffer value and fluid volume (V_{rel}) of the respective compartment, referred to 1kg of body water (for relevant references see Heisler, 1986a, 1986b)

Body compartment	Fishes						Mammals					
		Buffer value		Buffer capacity				Buffer value		Buffer capacity		
	V_{rel}	β_{NB}	β_{Bic}	κ_{NB}	κ_{Bic}		V_{rel}	β_{NB}	β_{Bic}	κ_{NB}	κ_{Bic}
Blood	~0.04	5-15	5-22	0.2-0.6	0.2-0.6		~0.08	20-30	40-50	1.6-2.4	3.2-4.0
ECS	~0.22	1-2	7-30	0.2-0.4	1.5-6		~0.20	2-4	50-60	0.4-0.8	10.0-12.0
ICS											
Skeletal muscle	~0.50	35-50	1-10	17.5-25	0.5-5		~0.35	65-78	15-20	22-27	5.2-7.0
Heart	~0.002	33-97	1-15	0.06-0.2	~0.02		~0.005	40-45	20-25	~0.2	0.1-0.2

β expressed as meq/(pH L body compartment water); κ expressed as meq/pH. ECS = extracellular space; ICS = intracellular space

Changes in P_{CO_2}

In higher vertebrates, changes in P_{CO_2} induced by changes in pulmonary ventilation are often utilised to compensate non-respiratory acid-base disturbances (cf. Woodbury 1965). Water-breathing species, however, are handicapped in applying this mechanism because of the lower oxygen content of water as compared with air (20-40 times, depending on temperature). Accordingly fishes have to ventilate their gas exchange systems at a 20- to 40-fold rate in order to maintain their oxygen supply at the same extraction ratio. Any further increase in ventilation of the comparatively viscous medium, water, would energetically be very uneconomical, and is only performed at increased oxygen demand. This may be related to the primarily oxygen-oriented regulation of ventilation in fishes (cf. Dejours 1975), which is certainly a limiting factor for the application of P_{CO_2} adjustments for fish acid-base regulation. Hyperventilation as a response to significantly lowered plasma pH is observed in fish only transiently (Randall, Heisler and Drees 1976), or when the primary ventilation stimulus of the low water oxygen content is eliminated by water hyperoxia (Heisler, Holeton and Toews 1981; cf. Heisler 1986b).

Hyperventilation is an inefficient mechanism for acid-base regulation in water-breathers anyway because of the small difference between plasma and water P_{CO_2} of only ~0.5 to ~3 mmHg (cf. Heisler 1984b, 1986b). This is the result of the much larger water solubility of carbon dioxide as compared with oxygen (35-23 times; 0 and 40°C, respectively) in combination with the high ventilation rate of water-breathing fishes (cf. Rahn 1966). Accordingly, P_{CO_2} is much lower than in air-breathers (usually <4 mmHg) and changes in environmental P_{CO_2}, which are small in absolute terms, induce relatively large changes in arterial P_{CO_2}, and much larger shifts in plasma pH. An increase in environmental P_{CO_2} by, for example, 8 mmHg will result in a considerable (~ five-fold) increase in plasma P_{CO_2} in fishes (typical control value: ~ 2 mmHg; cf. Heisler 1984b, 1986b), whereas the change would be only 20 per cent in a mammal (plasma P_{CO_2} ~ 40 mmHg). This rise in P_{CO_2} can easily be offset in mammals by an equivalent increase in alveolar ventilation, whereas in fishes even a tenfold increase in gill ventilation would not result in more than a 15 per cent reduction of the increased plasma P_{CO_2}. Elevation of inspired P_{CO_2} to the mentioned extent occurs frequently in the environment of fishes, but is rarely encountered by air-breathing species (cf. Heisler 1984b, 1986b).

Acid-Base Relevant Ion Transfer Processes

Transmembrane and transepithelial ion transfer processes are the only mechanisms capable of permanent elimination of H^+-equivalent ions produced as non-volatile metabolic end-products. These mechanisms are challenged by the steady-state metabolic production rate to only a small fraction of their capacity, but are taxed to their limits during extreme stress conditions (cf. Heisler 1984b, 1986b).

Transmembrane H^+-equivalent transfer is performed during acute and severe acid-base disturbances at a very high rate, which is comparable to the steady state diffusional carbon dioxide elimination (e.g. Benadé and Heisler 1978; Holeton and Heisler 1983; Holeton, Neumann and Heisler 1983). As a result of this high transmembrane transfer rate, the process of H^+ elimination from the tissue itself may become perfusion-limited (Neumann, Holeton and Heisler 1983). In contrast, when only small shifts in pH are to be readjusted, and adaptational processes are involved (e.g. temperature changes), the transfer may become rather slow (e.g. Heisler 1978).

The transepithelial ion transfer mechanisms are equipped with much smaller capacity than those for transmembrane transfer. This is not primarily related to the specific transfer capacity of the involved interfaces, but has to be attributed to the smaller surface area of the branchial epithelium compared with the interface area between cells and interstitial fluid. Even with this limitation, the transepithelial acid-base relevant transfer rate exhibited by fishes is much larger than that in higher vertebrates even in absolute terms (on a per kilogram basis), and larger by orders of magnitude when referred to the metabolic rate (Table 2.2).

Transepithelial transfer of HCO_3^-, OH^- or H^+ ions in the opposite direction has an identical effect for the acid-base status. Electroneutrality, however, can only be maintained by exchange with counter-ions of the same charge, or by co-transfer with an oppositely charged ion species. Co-transfer in excess to charged substances directly ingested, or produced in metabolism, is in conflict with the requirements of osmoregulation. Accordingly, acid-base relevant ionic transfer is most probably mainly performed as net 1:1 HCO_3^-/Cl^-, H^+/Na^+ and H^+/NH_4^+ ion exchange (for

Table 2.2: Maximal H[+] Equivalent Transepithelial Excretion Rates

Species	Conditions		Transfer rate ($\mu mol\,kg^{-1}$ body water)	Transfer rate/SMR	Reference
	Exercise-induced lactacidosis $[Lact^-]_{pl}$ (mM)	Hypercapnia P_{CO_2} (mmHg)			
Scyliorhinus stellaris	<15		14.3	0.41[a]	Holeton and Heisler, 1983
	>15		15.7	0.45[a]	
		8.2	15.3	0.43[a]	Heisler et al. 1976
		10.3	14.7	0.42[a]	Heisler and Neumann 1977
Conger conger	<10		9.3		Holeton et al. 1986
	10-20		10.4		
	>20		9.8		
Salmo gairdneri	<15		15.3		Toews et al. 1983
	>15	10.1	24.1	0.43[b]	Holeton et al. 1983
			26.3	0.47[b]	
Cyprinus carpio		37.2	5.3	0.13[c]	Claiborne and Heisler 1986
Bufo marinus		40.6	5.1	0.1[d]	Toews and Heisler 1982
Dog	Acute NH_4Cl acidosis		<3	0.02[e]	Pitts 1945a,b
	Chronic NH_4Cl acidosis		<7	0.03[e]	
Man	Severe diabetic acidosis		<7.5	0.03[e]	Pitts 1948

Values for standard metabolic rate (SMR):

[a] 34.8 $\mu mol\,min^{-1}\,kg^{-1}$ body water: Randall et al. 1976
[b] 55.3 $\mu mol\,min^{-1}\,kg^{-1}$ body water: Ultsch et al. 1980
[c] 39.1 $\mu mol\,min^{-1}\,kg^{-1}$ body water: Ultsch et al. 1980
[d] 55 $\mu mol\,min^{-1}\,kg^{-1}$ body water: Jackson and Braun 1979; data for bullfrog, Rana catesbeiana
[e] 298 $\mu mol\,min^{-1}\,kg^{-1}$ body water: Scotto et al. 1984

review see Maetz 1974; Evans 1979, 1980, 1984, 1986).

The exchange of ammonium ions against Na^+ does not appear to have a major role in physiological acid-base regulation, because the release of ammonia is hardly ever enhanced during stress conditions (cf. Heisler 1984b, 1986b). It has recently also been demonstrated that with normal environmental conditions ionic transfer of NH_4^+ is only insignificantly involved in the elimination of ammonia. By far the largest fraction is released by non-ionic diffusion similar to the mechanism for carbon dioxide (thus without any effect for the ionic regulation of the acid-base status). NH_4^+/Na^+ exchange becomes relevant only at environmental ammonia concentrations which are too high for non-ionic elimination of this nitrogenous waste product (Cameron and Heisler 1983, 1985; Evans 1986).

Correspondingly, HCO_3^-/Cl^- and H^+/Na^+ ion exchanges are mainly responsible for the observed branchial acid-base relevant transfer. Since all ion exchange mechanisms are considered as being carrier mediated, saturation of the carrier is a limiting factor for the time course of acid-base normalisation. Saturation occurs in intact fishes in the range of H^+ equivalent transfer rates between 5 and 25 µmol ($min^{-1}kg^{-1}$ body water) (Table 2.2). Ionic exchange processes are by definition sensitive to the availability of appropriate counterions, the lack of which must certainly be one of the potentially limiting factors of branchial acid-base relevant ion transfer.

According to their relatively high efficiency, transepithelial ion transfer processes play a much larger role in fishes than in higher vertebrates. They are not only responsible for steady state excretion of non-volatile metabolic end-products, but also provide transient acid-base regulation when other mechanisms are insufficient or too slow to eliminate the cause of disturbance by other means. Final adjustment of the acid-base status to permanent environmental changes is always performed via acid-base relevant ion transfer mechanisms.

Characteristics of Fish Acid-Base Regulation

pH as the central parameter of the acid-base status is quite tightly regulated even among different species, in spite of sometimes very different P_{CO_2} values (cf. Heisler 1984b, 1986b), This is attributable to the fine adjustment of the bicarbonate concentration by continuous transepithelial elimination of the steady state production of non-volatile metabolic end-products (cf. Heisler 1986a). When temperature varies, however, pH is generally reduced with rising temperature. The acid-base regulation during such changes in pH set-point values are in fishes mainly performed by transmembrane and transepithelial ion transfer processes supported by the negative temperature coefficients of the pK values of the physiological buffer systems. This topic cannot be adequately treated in this review, and the reader is referred to other relevant publications (e.g. Heisler 1984a, 1986a, b, c).

The stead state acid-base status cannot always be maintained, and deviations from the original pH in plasma and other body fluids cannot always be avoided during conditions of increased production of acidic metabolic end products, or changes in P_{CO_2} in the body fluids brought about by ambient hypercapnia, or hypercapnia induced by changes in convective gas exchange.

Lactacidosis

Activity of the poorly perfused white musculature often results in energy production with lactic acid as metabolic end product of anaerobic glycolysis. Immediately upon production lactic acid is dissociated producing lactate$^-$ and H$^+$ ions, which affect acid-base disturbances in the muscle fibres and, after release to the extracellular space, also in other body compartments. These disturbances are characterised by a sharp drop in plasma pH and [HCO$_3^-$], and a rise in P_{CO_2}, with peak deflections from control values within 1 h after the muscular activity. Peak plasma lactate concentrations, however, are not attained before 2-8 h after anaerobic muscular activity (Figure 2.1).

Further aerobic processing of lactic acid (synthesis to glycogen or oxidation to carbon dioxide) mainly in liver, heart muscle and red muscle requires up to 30 h depending on the magnitude of the imposed stress. The plasma acid-base status is restored much earlier (Figure 2.1). Plasma [HCO$_3^-$] and plasma pH are normalised at latest after about 12 h (pH in some instances after 18 h). Information on the time course of intracellular pH restoration is sparse, but intracellular pH (pH$_i$) is likely to be restored as well when the extracellular pH has achieved control conditions. The initial deflection of intracellular muscle pH in the elasmobranch *Scyliorhinus stellaris* is larger than that of arterial and

Figure 2.1: Characteristic Changes in Plasma Acid-Base Parameters, Plasma Lactate Concentration, and Transfer of H^+-equivalent Ions between Extracellular Space and Environmental Water ($\Delta H^+_{e \to w}$) after Strenuous Muscular Activity. (For values and times of maximal deflection (index 'm') and return to control values (index 'c') refer to Heisler 1984b, 1986b)

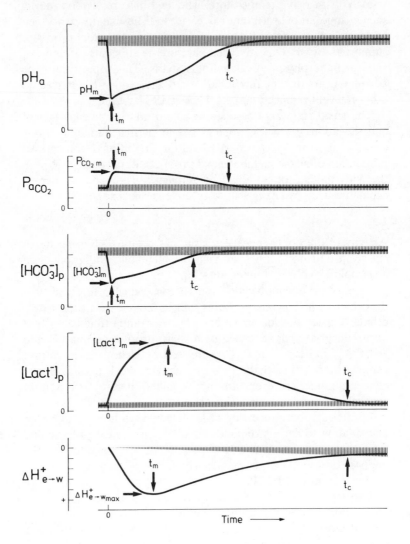

venous plasma pH, but muscle pH_i recovers towards control values much faster than plasma pH (Figure 2.2) (for details and references see Heisler 1984b, 1986b).

Recovery of the acid-base status before removal of lactate is the result of a considerable transfer of acid-base relevant ions to the environmental water (Figure 2.1). In *Scyliorhinus*, for instance, about 45 per cent of the originally produced H^+ ions are shifted within about 10 h into the environmental water, which at that time represents more than 80 per cent of the surplus H^+ ions not yet returned together with lactate⁻ into metabolism (Holeton and Heisler 1983). During the following period of the recovery phase, H^+-equivalent ions are taken back up from the water in order to avoid alkalinisation by metabolic processing of the remaining lactic acid (Figure 2.2).

This pattern of transient branchial transfer of surplus H^+ ion equivalents to the environment facilitates normalisation of the acid-base status long before the original stress factor, lactic acid, has been completely removed. A second important feature of the regulatory pattern is the preferential restoration of intracellular pH, suggesting priority of intracellular over extracellular pH regulation.

Hypercapnia

Hypercapnia in fishes may be induced by various factors. The most obvious one is an increase in environmental P_{CO_2}, which occurs in natural habitats due to hindrance of gas exchange between water and air by dense water surface vegetation and by thermo-stratification (cf. Heisler 1984b, 1986b). In response to ambient hypercapnia, which is transmitted to the plasma almost immediately via the large branchial surface area, arterial pH is reduced due to the elevation of plasma P_{CO_2} (Figure 2.3). Plasma pH recovers towards control values and is restored to usually less than -0.1 units by elevation of the plasma $[HCO_3^-]$. The relatively fast increase of plasma bicarbonate during the initial phase of recovery is related to the supply of bicarbonate by intracellular body compartments. During this period bicarbonate is not yet gained, but to a limited extent is even lost to the environment (cf. Figure 2.3 lower panel). Uptake of bicarbonate equivalents from (or excretion of net H^+ equivalents to) the water is initiated with a delay of up to half an hour. The amount of bicarbonate originally supplied by the intracellular body compartments is then returned.

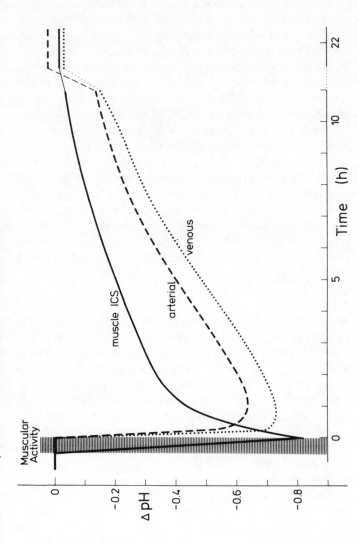

Figure 2.2: Deflection of Arterial and Venous Plasma pH, and of Intracellular (ICS) Muscle pH as a Function of Time after Strenuous Muscular Activity (based on data of Holeton and Heisler 1983)

Figure 2.3: Characteristic Behaviour of Arterial Plasma pH, P_{CO_2} and [HCO_3^-], and H^+-equivalent Ion Transfer between Extracellular Space and Environmental Water during Exposure of Fish to Environmental Hypercapnia

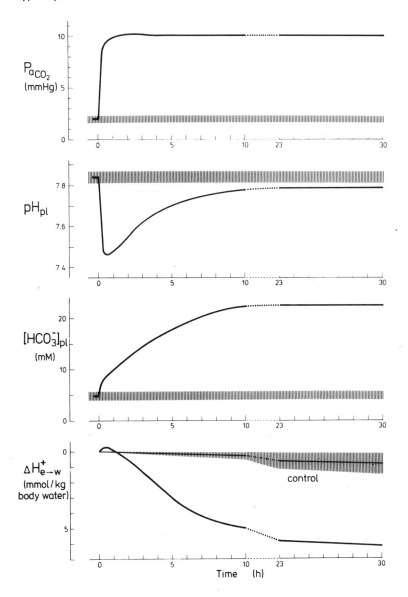

Additional bicarbonate transferred to the intracellular compartments finally results in even more complete intracellular pH compensation (ΔpH_i < 0.05) than in the extracellular space (see Heisler 1984b, 1986b).

Hyperoxia in the environmental water regularly results in a reduction of gill ventilation due to the primarily oxygen-oriented regulation of gill ventilation (cf. Dejours 1975), which effects an increase in arterial P_{CO_2} (see Heisler 1986b). Although the rise in P_{CO_2} is slower than during exposure to environmental hypercapnia, the degree of plasma pH compensation is usually similar when the new steady state P_{CO_2} has been attained. An interesting feature of the regulatory pattern in the elasmobranch *Scyliorhinus stellaris* is that the reduction in gill ventilation is matched to the rate of uptake of HCO_3^--equivalent ions from the water, such that plasma pH is reduced from control values not more than -0.08 pH units during the whole process of adaptation to the new steady state levels (for most of the time <-0.03; Figure 2.7, cf. Heisler 1984b, 1986b).

The switch from water breathing to air breathing in facultative air-breathing fishes is comparable to exposure of water-breathing fishes to environmental hyperoxia. In both cases the oxygen content of the gas exchange medium is increased considerably. Accordingly P_{CO_2} is elevated also in facultative air-breathers (e.g. 5.6 to 26 mmHg within 2-3 days in *Synbranchus*; Heisler 1982). In contrast to water breathers, however, air-breathing fishes usually do not compensate the associated fall in plasma pH by an active increase of the plasma bicarbonate concentration (cf. Heisler 1984b, 1986b), although intracellular pH in heart and white muscle tissues of *Synbranchus* is very well compensated. The bicarbonate accumulated in the tissues is produced by carbon dioxide buffering by the haemoglobin-rich blood of *Synbranchus* and shifted into the intracellular tissue compartments elevating bicarbonate in surplus to the amount produced by intracellular non-bicarbonate buffering (Heisler 1982). This example of an extreme protection of the intracellular body compartments confirms the idea of preferential regulation of intracellular *vs.* extracellular pH regulation, which is also supported by the data obtained during environmental hypercapnia, and during recovery from lactacidosis.

Limitations of Acid-Base Regulation

The effectiveness of individual mechanisms for acid-base regulation has to be evaluated in context with the border conditions. Normalisation of pH in a certain body compartment is dependent on a variety of processes which may not be limiting *per se*, but may become limiting in combination with features of other mechanisms. A typical example for such an interaction can be demonstrated for the elimination of the dissociation products of lactic acid from the muscle cells.

Lactacidosis

The acid-base status immediately after anaerobic muscular activity is characterised by an extremely inhomogeneous distribution of surplus H^+ ions among various body compartments (cf. Figure 2.2). Since white muscle tissue is incapable of significant further aerobic processing of lactic acid, the dissociation products of lactic acid have to be eliminated from the site of production. This process is governed to only a very limited extent by the efflux time constants of H^+ and lactate ions. In spite of the fact that the time constant for H^+ is larger than that for lactate by at least one order of magnitude (e.g. Benadé and Heisler 1978; Holeton and Heisler 1983), the amount of lactate present in the extracellular space usually exceeds the amount of surplus H^+ ions considerably. This apparent discrepancy, termed 'H$^+$ ion deficit' (Piiper, Meyer and Drees 1972), is the result of a number of factors interacting in a relatively complex way.

Lactate is transferred from the intracellular body compartments to the extracellular space (ECS), from which it is removed by metabolic processes in liver, heart muscle and also red muscle tissue, but is never released to the environment. Since this process is presumably performed by passive diffusion, lactate is distributed across the cell membrane either according to the membrane potential (main transfer form: lactate ions) or according to the intracellular/extracellular pH difference (main transfer form: lactic acid). In both cases, equilibrium is achieved when the majority of lactate has been transferred to the ECS. The time course of transfer is related to the diffusional efflux curve, and the overlying relatively constant rate of metabolism.

In contrast to lactate, the distribution pattern of surplus H^+ ions is mainly governed by the distribution of the buffer capacity, which in fishes is higher in the intracellular muscle compartments than in the ECS by a factor in the range of 5 to 100 (cf. Table 2.1). Accordingly, transfer of only a small amount of H^+ ions is sufficient to lower extracellular pH so far that a new equilibrium is attained between intra- and extracellular pH which eliminates the driving force for any further transfer of H^+ ions to the extracellular space. This 'equilibrium limitation' (Holeton *et al.* 1983) implies that only a small fraction of the surplus H^+ ions can initially be transferred to the ECS (Figure 2.4). With this situation the bulk of surplus H^+ ions would be left buffered in the intracellular body compartments until further metabolic processing of lactic acid would occur, unless surplus H^+ ions were transiently passed to the ambient water to a considerable extent (Figure 2.4). Accordingly the rate of net H^+ transfer from intracellular to extracellular space is primarily limited to the combined rate of H^+ removal from the ECS by metabolisation of lactic acid and net H^+ transfer to the ambient water.

The following quantitative analysis based on data collected for the elasmobranch fish *Scyliorhinus stellaris* is certainly not generally valid for different conditions and species, but should be taken as an exemplary demonstration of the time-course limiting factors for acid-base normalisation during lactacidosis.

After termination of muscular activity about $16\,mmol\,kg^{-1}$ body water of lactic acid is accumulated in intracellular body compartments (Figure 2.4). H^+ ions are released from the muscle cells at an extremely high rate, expressed by the tangent to the initial part of the ΔH^+_i curve ($> 150\ \mu mol\,min^{-1}kg^{-1}$ body water, S_1, Figure 2.4). The initial rise in the curve of extracellularly buffered H^+ ions (Figure 2.4, S_3) is smaller than the efflux rate from the ICS by only the small rate of lactic acid metabolism (S_4). Although the value for this initial H^+ efflux from the muscle cells (S_1) appears to be extremely large, it nevertheless represents an underestimate of the real transmembrane transfer rate. The elimination process is most likely perfusion limited (Neumann *et al.* 1983). Furthermore, the transfer of H^+ equivalents into non-muscular intracellular compartments, facilitated by reduction of extracellular pH, will also result in an apparent reduction of S_1. This factor can hardly be quantified on the basis of the available data.

After an initial very short time period of less than one hour, the high H^+-equivalent ICS to ECS transfer rate is considerably

Figure 2.4: Distribution of H⁺ and Lactate Ions among Intracellular ('i') and Extracellular ('e') Body Compartments, and the Environmental Water ('e→sw') after Strenuous Muscular Activity in *Scyliorhinus stellaris*. S_1 is the initial net intracellular to extracellular H⁺ transfer rate, S_2 is the difference between the rate of metabolic processing of lactic acid (S_4) and the rate of net H⁺ transfer to the environmental water (S_5). S_3 is the initial slope of occurrence of surplus H⁺ ions in the extracellular compartment, equivalent to the difference between S_1 and S_4. For details see text

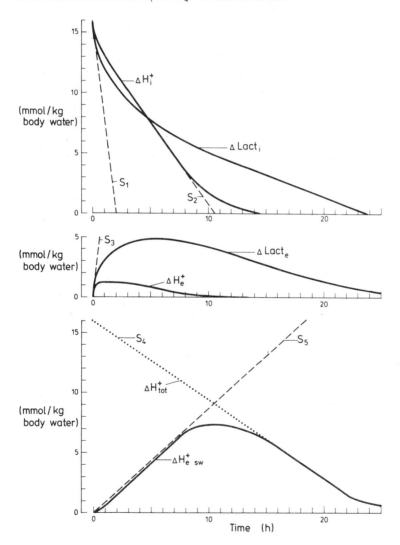

reduced to about $25\,\mu mol\,min^{-1}kg^{-1}$ (Figure 2.4, S_2) mainly as a result of the 'equilibrium limitation'. Extracellular pH in the small volume of poorly buffered extracellular fluid is reduced by transfer of only about 8 per cent of the amount of originally produced H^+ ions to such an extent that equilibrium (though at a reduced level) between intracellular and extracellular pH is attained. This equilibrium eliminates any further H^+-equivalent transfer in excess of the rate of H^+ removal from the ECS by other mechanisms. After having attained pH equilibrium, the time course of pH normalisation is therefore only dependent on the two factors of lactic acid metabolism (~ 11 μmol $min^{-1}kg^{-1}$: S_4) and the transfer rate of H^+ equivalents to the environmental water (~ 14 μmol $min^{-1}kg^{-1}$, S_5). When all surplus H^+ ions are removed from the organism, metabolism of lactic acid continues with H^+ ions being taken back up from the environment at a rate determined by overall metabolism (S_4, ~ 11 $\mu mol\,min^{-1}kg^{-1}$), which is less than the maximal branchial transfer rate (S_5, ~ 14 $\mu mol\,min^{-1}kg^1$).

Hypercapnia

In contrast to lactacidosis, during hypercapnia H^+ ions are not produced exclusively in certain body compartments, but hydration and dissociation of carbon dioxide takes place more uniformly distributed within the body fluids. The induced changes in extracellular pH, however, are larger than those of pH in intracellular body compartments due to the different non-bicarbonate buffer values (cf. Table 2.1). The disturbed equilibrium between intracellular and extracellular pH is quickly re-established on a different level by transmembrane transfer of H^+, or HCO_3^--equivalent ions, a mechanism contributing significantly to the initial compensation of the extracellular pH during environmental hypercapnia (cf. Heisler, Weitz and Weitz 1976; Toews, Holeton and Heisler 1983).

Any further increase in plasma bicarbonate is exclusively due to uptake of bicarbonate-equivalent ions from the environmental water, the rate of which, however, is extremely variable. With some exceptions complete compensation ($\Delta pH < -0.1$) is attained, although the time required for this process differs by orders of magnitude among species and studies. Differences in the composition of the environmental water may at least partially be responsible for these discrepancies. The time course of compensation is similar in the marine fish *Conger conger,* the channel catfish *Ictalurus punctatus,* and the fresh-water rainbow trout *Salmo gairdneri,* all three kept at relatively high pH and [HCO_3^-], whereas *Salmo gairdneri* and *Cyprinus carpio* do not achieve compensation at relatively low water pH and [HCO_3-] before exposure to hypercapnia for an 8- to 40-fold time period (Figure 2.5 and Table 2.3).

Systematic variation of the water bicarbonate concentration affected the time course of hypercapnia compensation in the elasmobranch fish *Scyliorhinus stellaris* to a similar extent. The compensatory bicarbonate uptake by this species was speeded up at increased sea-water pH (via bicarbonate elevation), whereas reduction of sea-water pH to 6.1 eliminated any compensatory bicarbonate-equivalent ion exchange completely. The rate of bicarbonate-equivalent ionic transfer was maximal and constant in a range of -0.5 to 0.6 of the difference between plasma and sea-water pH ($pH_{pl} - pH_{sw}$), decreased linearly at higher values reaching zero at $pH_{pl} - pH_{sw} = 1.1$, and was reversed into a HCO_3^--equivalent ion loss at lower water pH values (Figure 2.6). The essentially constant water [Na^+] (~480 mM) suggested the environmental bicarbonate concentration as the limiting factor for acid-base relevant branchial ion exchange (Heisler and Neumann 1977).

These data shed light on the importance of appropriate ion concentrations in the environmental water for the acid-base regulation in fishes. Since air-breathing (aestivating) fishes usually have access to only a very limited water volume, which furthermore is soon depleted of bicarbonate, or electrolyte poor *a priori* (cf. Heisler 1982), non-compensation of the developing hypercapnia could well be related to the lack of environmental ions. Other factors, however, are more likely to be responsible for the observed phenomenon. An inherent feature of aerial gas exchange in most air-breathers is the reduction of the fractional contact time between water and gill epithelium — which is considered as the main ionic transfer site in these animals also — by factors of 100 and more as compared to water breathing (Heisler 1982). It is evident that such conditions allow hardly any significant ion exchange process.

The principle of P_{CO_2} elevation during hyperoxia is similar to that for air-breathing fishes: reduced ventilation induced by the comparatively higher oxygen content of the gas exchange medium

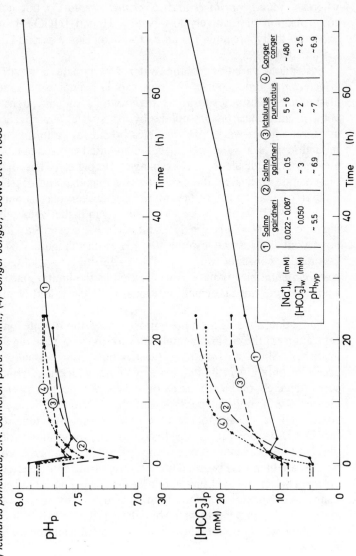

Figure 2.5: Time Course of pH Compensation by Extracellular Bicarbonate Accumulation Compared with the Environmental Water Composition in Three Fish Species. (1) *Salmo gairdneri*, Janssen and Randall 1975; (2) *Salmo gairdneri*, Eddy *et al.* 1977; (3) *Ictalurus punctatus*, J.N. Cameron pers. comm.; (4) *Conger conger*, Toews *et al.* 1983

Table 2.3: Plasma Bicarbonate Concentrations during Control Conditions ('c'), after Exposure to Hypercapnia for Variable Time Periods ('hyp'), and Extracellular pH Compensation in Fishes during Hypercapnia

Species	Cause of hypercapnia	$[HCO_3^-]_c$ (mM)	$[HCO_3^-]_{hyp}$ (mM)	pH comp	Exposure time (h)	Reference
Scyliorhinus stellaris	Ambient	7.6	20	85	8	Heisler et al. (1976)
		6	19	85	4	Randall et al. (1976)
Scyliorhinus stellaris	Hyperoxia-induced	5.3	20	95	25	Heisler et al. (1981, 1986)
		5.3	24	90	144	
Squalus acanthias	Ambient	8	21	64	5	Cross et al. (1969)
Conger conger	Ambient	5.0	22	90	10 }	Toews et al. (1983)
			22	90	30 }	
Ictalurus punctatus	Ambient	7	21	90	30	Cameron (1980)
Cyprinus carpio	Ambint	13.8	22	40	48	Claiborne and Heisler (1984)
		11.4	28	45	96 }	Claiborne and Heisler (1986)
		13.0	26	80	456 }	
		12.0	23	55	120	
Cyprinus carpio	Hyperoxia-induced	12.5	21	40	72	N.A. Anderson, J.B. Claiborne and N. Heisler, unpubl.
Salmo gairdneri	Ambient	9	27	95	75	Janssen and Randall (1975)
		3.8	23	90	>168	Eddy et al. (1977)
	Hyperoxia-induced	7.0	25.8		72	Höbe et al. (1984)
Synbranchus marmoratus	Transition, water- to air-breathing	24	24	0	18 }	Heisler (1982)
		24	24	0	100 }	

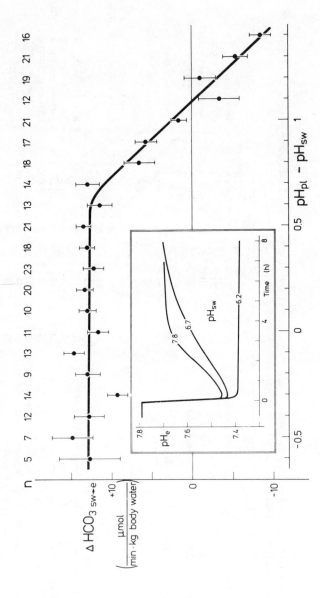

Figure 2.6: Rate of Hypercapnia-induced Bicarbonate-equivalent Ion Transfer between Sea-Water and Extracellular Space as a Function of the Plasma-Sea Water pH Difference ($pH_{pl} - pH_{sw}$). Insert: time course of plasma pH recovery as a function of sea-water pH (pH_{sw} modified by changes in bicarbonate concentration)

leads to a rise in P_{CO_2}. In contrast to air-breathing, however, contact between the branchial ion-exchange epithelium and the ambient water is maintained, facilitating acid-base relevant ionic transfer. Accordingly, the reduction in plasma pH accompanying the rise in P_{CO_2} is usually compensated at least to a limited extent.

Data collected during hyperoxia-induced hypercapnia in *Scyliorhinus* reveal another limiting factor for pH compensation. The rate of bicarbonate gain from the environment is initially similar to those observed during lactacidosis and environmental hypercapnia, but is reduced when plasma pH is compensated to $\Delta pH < -0.03$. Bicarbonate uptake continues in parallel with the rise in P_{CO_2}, but is considerably reduced at plasma bicarbonate of higher than 20 mM. Any bicarbonate gain is essentially terminated at an $[HCO_3^-]$ of more than 22 mM (< 0.2 $\mu mol\,min^{-1}kg^{-1}$). P_{CO_2}, however, continues to rise, leading to progressive decompensation of plasma pH ($\Delta pH > 0.01$, (Figure 2.7) (cf. Heisler 1984b, 1986b).

The bicarbonate level of 24-25 mM attained at partial decompensation of plasma pH after 6 days of environmental hypercapnia appears to be indicative of the maximal concentration that can be achieved by the bicarbonate-retaining and bicarbonate-resorbing structures in fishes. This level is hardly ever surpassed during hypercapnia of various origin independent of the time length of hypercapnic exposure (Table 2.3), and may explain the extreme discrepancy in the extent of compensation observed between the majority of fish species and the 'exceptions to the rule', namely *Cyprinus* and *Synbranchus* (Table 2.3). It becomes evident that the extent of compensation is related to the control bicarbonate concentration. The higher the initial bicarbonate concentration (carp: 13 mM, *Synbranchus*: 24 mM, as compared with 3-8 mM of other fishes, cf. Heisler 1984b, 1986b), the smaller is the possible relative compensatory increase during hypercapnia and thus restoration of pH towards control values (Heisler 1986c).

The maximal plasma $[HCO_3^-]$ in individual species is not influenced by hypercapnia of larger extent (5 per cent), or by increased levels of water electrolyte concentrations (Claiborne and Heisler 1986). Even infusion of bicarbonate (5 mmol kg^{-1} body weight) only transiently increased plasma $[HCO_3^-]$, before the administered amount was quantitatively excreted into the environmental water (Claiborne and Heisler 1986). Accordingly, a concentration

Figure 2.7: Compensation of Plasma pH during Hyperoxia-induced Hypercapnia. Plasma pH is kept essentially constant by elevation of plasma $[HCO_3^-]$ to about 22 mM. Further increase in P_{CO_2} results in only slight further $[HCO_3^-]$ increase and leads to partial pH decompensation. See text for details. (Data of Heisler *et al.* 1981.)

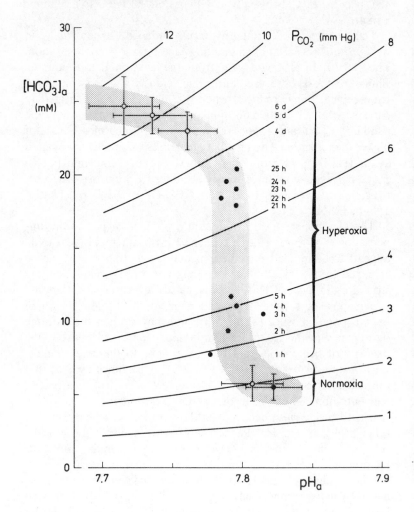

of about 25 mM has to be considered as bicarbonate threshold in plasma and as a constitutional limit in at least some fish species.

Conclusion

Acid-base regulation in fishes generally aims for normalisation of pH before the original stress factor is removed, and intracellular pH is preferentially regulated as compared with extracellular pH. In contrast to terrestrial animals, this task is supported to only a very limited extent by changes in P_{CO_2} and by buffering processes. The major proportion and the final adjustment of the acid-base status is instead performed by transmembrane and transepithelial bicarbonate-equivalent ion transfer processes.

The performance of the ion exchange mechanisms involved is influenced by various constitutional, behavioural and environmental factors. The capacity of transmembrane acid-base relevant transfer mechanisms is large, and is *in vivo* only limited by tissue perfusion, and by the H^+ ion capacitance of tissue perfusate and total extracellular space. The branchial ion transfer mechanisms are limited by carrier-related transport maxima, by reduced relative contact time of the ion exchange epithelium with water, by the plasma bicarbonate threshold, and by the availability of environmental ions.

These factors contribute to the delay of acid-base normalisation to an extent which is variable with the type of acid-base stress. The major proportion is usually attributable to the transepithelial ion transfer. The sensitivity of the involved mechanisms to environmental ion concentrations suggests that this factor is responsible for the wide variability of the time required for pH normalisation reported in the literature.

References

Benadé, A.J.S. and Heisler, N. (1978) 'Comparison of Efflux Rates of Hydrogen and Lactate Ions from Isolated Muscles *in vitro*', *Respiration Physiology, 32*, 369-80

Cameron, J.N. (1980) 'Body Fluid Pools, Kidney Function, and Acid-Base Regulation in the Freshwater Catfish *Ictalurus punctatus*', *Journal of Experimental Biology, 86*, 171-85

Cameron, J.N. and Heisler, N. (1983) 'Studies of Ammonia in Rainbow Trout: Physico-chemical Parameters, Acid-Base Behaviour and Respiratory

Clearance', *Journal of Experimental Biology, 105*, 107-25

Cameron, J.N. and Heisler, N. (1985) 'Ammonia Transfer Across Fish Gills: a Review', in R. Gilles (ed.), *Circulation, Respiration and Metabolism*, Springer, Heidelberg, pp. 91-100

Claiborne, J.B. and Heisler, N. (1984) 'Acid-Base Regulation in the Carp (*Cyprinus carpio*) during and after Exposure to Environmental Hypercapnia', *Journal of Experimental Biology, 108*, 25-43

Claiborne, J.B. and Heisler, N. ('1986). Acid-Base Regulation and Ion Transfers in the Carp (*Cyprinus carpio*): pH Compensation during Graded Long- and Short-term Environmental Hypercapnia and the Effect of Bicarbonate Infusion', *Journal of Experimental Biology*, in press.

Cross, C.E., Packer, B.S., Linta, J.M., Murdaugh, H.V., Jr and Robin, E.D. (1969) 'H⁺ Buffering and Excretion in Response to Acute Hypercapnia in the Dogfish *Squalus acanthias*', *American Journal of Physiology, 216*, 440-52

Dejours, P. (1975) *Principles of Comparative Respiratory Physiology*, North-Holland, Amsterdam

Eddy, F.B., Lomholt, J.P., Weber, R.E. and Johansen, K. (1977) 'Blood Respiratory Properties of Rainbow Trout (*Salmo gairdneri*) Kept in Water of High CO_2 Tension', *Journal of Experimental Biology, 67*, 37-47

Evans, D.H. (1979) 'Fish' in G.M.O. Maloiy (ed.), *Comparative Physiology of Osmoregulation in Animals*, vol. I, Academic Press, New York, pp. 305-90

Evans, D.H. (1980) 'Kinetic Studies of Ion Transport by Fish Gill Epithelium', *American Journal of Physiology, 238*, R224-30

Evans, D.H. (1984) 'The Role of Gill Permeability and Transport Mechanisms in Euryhalinity', in W.S. Hoar and D.J. Randall (eds), *Fish Physiology*, vol. XB, Academic Press, Orlando, pp. 315-401

Evans, D.H. (1986) 'The Role of Branchial and Dermal Epithelia in Acid-Base Regulation in Aquatic Animals', in N. Heisler (ed), *Acid-Base Regulation in Animals*, Elsevier Biomedical Press, Amsterdam, pp. 139-72

Heisler, N. (1978) 'Bicarbonate Exchange between Body Compartments after Changes of Temperature in the Larger Spotted Dogfish (*Scyliorhinus stellaris*)', *Respiration Physiology, 33*, 145-60

Heisler, N. (1982) 'Intracellular and Extracellular Acid-Base Regulation in the Tropical Freshwater Teleost Fish *Synbranchus marmoratus* in Response to the Transition from Water Breathing to Air Breathing', *Journal of Experimental Biology, 99*, 9-28

Heisler, N. (1984a) 'Role of Ion Transfer Processes in Acid-Base Regulation with Changes of Temperature in Fish', *American Journal of Physiology, 246*, R551

Heisler, N. (1984b) 'Acid-Base Regulation in Fishes', in W.S. Hoar and D.J. Randall (eds), *Fish Physiology*, vol. XA, Academic Press, Orlando, pp. 315-401

Heisler, N. (1986a) 'Buffering and Transmembrane Ion Transfer Processes', in N. Heisler (ed.), *Acid-Base Regulation in Animals*, Elsevier Biomedical Press, Amsterdam, pp. 3-47

Heisler, N. (1986b) 'Acid-Base Regulation in Fishes', in N. Heisler (ed.), *Acid-Base Regulation in Animals*, Elsevier Biomedical Press, Amsterdam, pp. 309-56

Heisler, N. (1986c) 'Comparative Aspects of Acid-Base Regulation', in N. Heisler (ed.), *Acid-Base Regulation in Animals*, Elsevier Biomedical Press, Amsterdam, pp. 397-450

Heisler, N. and Neumann, P. (1977) 'Influence of Sea Water pH upon Bicarbonate Uptake Induced by Hypercapnia in an Elasmobranch (*Scyliorhinus stellaris*)', *Pflügers Archiv, 368*, Suppl., R19

Heisler, N., Weitz, H. and Weitz, A.M. (1976) 'Hypercapnia and Resultant

Bicarbonate Transfer Processes in an Elasmobranch Fish', *Bulletin Européen de Physiopathologie Respiratoire, 12*, 77-85

Heisler, N., Holeton, G.F. and Toews, D.P (1981) 'Regulation of Gill Ventilation and Acid-Base Status in Hyperoxia-induced Hypercapnia in the Larger Spotted Dogfish (*Scyliorhinus stellaris*)', *Physiologist, 24*, 305

Höbe, H., Wood, C.M. and Wheatly, M.G. (1984) 'The Mechanisms of Acid-Base and Ionoregulation in the Freshwater Rainbow Trout during Environmental Hyperoxia and Subsequent Normoxia. I. Extra- and Intracellular Acid-Base Status', *Respiration Physiology, 55*, 139-54

Holeton, G.F. and Heisler, N. (1983) 'Contribution of Net Ion Transfer Mechanisms to the Acid-Base Regulation after Exhausting Activity in the Larger Spotted Dogfish (*Scyliorhinus stellaris*)', *Journal of Experimental Biology, 103*, 31-46

Holeton, G.F., Neumann, P. and Heisler, N. (1983) 'Branchial Ion Exchange and Acid-Base Regulation after Strenuous Exercise in Rainbow Trout (*Salmo gairdneri*)' *Respiration Physiology, 51*, 303-18

Jackson, D.C. and Braun, B.A. (1979) 'Respiratory Control in Bullfrogs: Cutaneous Versus Pulmonary Response to Selective CO_2 Exposure', *Journal of Comparative Physiology, 129B*, 339-42

Janssen, R.G. and Randall, D.J. (1975) 'The Effect of Changes in pH and P_{CO_2} in Blood and Water on Breathing in Rainbow Trout, *Salmo gairdneri*', *Respiration Physiology, 25*, 235-45

Maetz, J. (1974) 'Adaptation to Hyper-osmotic Environments', *Biochemical and Biophysical Perspectives in Marine Biology, 1*, 91-149

Neumann, P., Holeton, G.F. and Heisler, N. (1983) 'Cardiac Output and Regional Blood Flow in Gills and Muscles after Exhaustive Exercise in Rainbow Trout (*Salmo gairdneri*), *Journal of Experimental Biology, 105*, 1-14

Piiper, J., Meyer, M. and Drees, F. (1972) 'Hydrogen Ion Balance in the Elasmobranch *Scyliorhinus stellaris* after Exhausting Activity', *Respiration Physiology, 16*, 290-303

Pitts, R.F. (1945a and b) 'The Renal Regulation of Acid-Base Balance with Special Reference to the Mechanism of Acidifying the Urine', *Science 102*, 49-54 and 81-5

Pitts, R.F. (1948) 'Renal Excretion of Acid', *Federation Proceedings, 7*, 418-26

Rahn, H. (1966) 'Aquatic Gas Exchange: Theory', *Respiration Physiology, 1*, 1-12

Randall, D.J., Heisler, N. and Drees, F. (1976) 'Ventilatory Response to Hypercapnia in the Larger Spotted Dogfish *Scyliorhinus stellaris*', *American Journal of Physiology, 230*, 590-4

Scotto, P., Rieke, H., Schmitt, H.J., Meyer, M. and Piiper, J. (1984) 'Blood-Gas Equilibration of CO_2 and O_2 in Lungs of Awake Dogs during Prolonged Rebreathing', *Journal of Applied Physiology: Respiratory Environmental Exercise Physiology 57*, 1354-9

Toews, D.P. and Heisler, N. (1982) 'The Effects of Hypercapnia on Intracellular and Extracellular Acid-Base Status in the Toad *Bufo marinus*', *Journal of Experimental Biology, 97*, 79-86

Toews, D.P., Holeton, G.F. and Heisler, N. (1983) 'Regulation of the Acid-Base Status during Environmental Hypercapnia in the Marine Teleost Fish *Conger conger*', *Journal of Experimental Biology, 107*, 9-20

Ultsch, G.R., Ott, M.E. and Heisler, N. (1980) 'Standard Metabolic Rate, Critical Oxygen Tension, and Aerobic Scope for Spontaneous Activity of Trout (*Salmo gairdneri*) and Carp (*Cyprinus carpio*) in Acidified Water', *Comparative Biochemistry and Physiology, 67A*, 329-35

Woodbury, J.W. (1965) 'Regulation of pH', in T.C. Ruch and H.D. Patton (eds), *Physiology and Biophysics*, Saunders, Philadelphia, pp. 899-934

3 PHYSIOLOGICAL INVESTIGATIONS OF MARLIN

Charles Daxboeck and Peter S. Davie

There are four species of marlin (Istiophoridae): blue (*Makaira nigricans*), black (*M. indica*), striped (*Tetrapterus audux*) and white (*T. albidus*). All are large, pelagic, predatory teleosts that range the warm temperate and tropical oceans. Indeed it is presently thought that the Pacific blue and black marlins are the largest extant teleosts, reportedly exceeding 1200 kg based on unconfirmed Japanese commercial longline data. A single specimen of a blue marlin weighing 820.5 kg (1805 lb) is the largest confirmed capture on rod and reel (Mather 1976). The ability of marlin to perform short-term explosive burst swimming is legendary, and they can attain estimated maximum speeds of 40 to $60 \mathrm{km\,h^{-1}}$ (see Magnuson 1978 for data on other scombrids). They also have rather prodigious capacities for steady-state swimming and are obligate ram ventilators. Based on recent tag recoveries (NMFS News Release, 11 January 1985) a single black marlin travelled a record 10618 km, from Baja California to Norfolk Island in the south-west Pacific off New Zealand, at an average straight-line speed of $17.3 \mathrm{km\,day^{-1}}$.

In this chapter we shall examine some of the possible physiological mechanisms by which these magnificent fish are able to accomplish their quite incredible feats of exercise activity. Because of their large size and predatory behaviour, direct measurements of *in vivo* parameters are not presently available to investigators because of severe logistic constraints. Given these limitations there have none the less been several recent studies concerning various aspects of marlin physiology. These include studies of thermogenic tissues in eye muscles ('brain heaters'; Block 1983, 1984), blood chemistry and acid-base regulation following exercise (Wells and Davie 1985; Dobson, Wood, Daxboeck and Perry 1985), muscle biochemistry and biomechanics (Johnston and Salamonski 1984; Altringham and Johnston 1985; Johnston and Altringham 1985; Suarez Mallet, Daxboeck and Hochachka 1985) and coronary vasculature function (Davie and Daxboeck 1984). Selected topics will be discussed in some detail in the following sections.

Table 3.1: Some Respiratory and Acid-Base Properties of the Blood of Pacific Blue (*M. nigricans*) and Striped (*T. audax*) Marlin at Time of Capture. (Data from Dobson *et al.* (1985) for blue marlin and from Wells and Davie (1985) for striped marlin.)

	Blue (25°C)	Striped (25°C)
Weight (kg)	94.5 ± 5.4 (29)	67-172 (6)
Haemoglobin	10.4 ± 0.4 (g 100 ml⁻¹) (29)	1.20 ± 0.05 (mmol l⁻¹) (6)
Haematocrit (%)	43.1 ± 1.7 (26)	24.2 ± 1.35 (6)
pH whole blood	7.22 ± 0.07 (13)	6.7 ± 0.03 (6) (no CO₂)
pH$_i$	7.04 ± 0.06 (13)	n.a.
P_{O_2}	9.7 ± 3.0 (8)	n.a.
P_{aCO_2}	17.3 ± 3.1 (8)	n.a.
Total CO₂ (mM)	9.5 ± 0.1 (16)	n.a.
Lactate (mM)	12.8 ± 2.1 (15)	15.03 ± 0.27 (6)
Whole blood buffer value		
$\dfrac{dHCO_3^-}{dpH}$	21.3 Slykes	n.a.
Bohr coefficient	−1.00 (9)	−0.74 (6)
(φ, whole blood)	(pH 6.95-7.60)	(pH 6.6-8.1)

Numbers represent means ± SEM; those in parentheses represent number of fish sampled.

Blood

Recent Findings

Blood acid-base and respiratory properties for marlin represent data from fish in a highly specific physiological and metabolic state: that of an animal that has undergone severe exercise stress. Since it is not yet possible to determine the parameters presented in Table 3.1 from samples taken from free-swimming marlin in a resting state, they can only be assessed against comparable values for other teleosts (see Jones and Randall 1978). At the time of capture, marlin blood pH is very acidic by teleost standards, assuming a pre-exercise value of about pH 7.80 at 25°C. There are two potential sources for the noted increase in hydrogen ions (H⁺) in the blood; metabolic (aerobic and/or anaerobic) and respiratory.

From the respiratory aspect, the reduction in blood pH following extreme exercise is accompanied by an increased P_aCO_2, of around 17 Torr and a decreased P_aCO_2 of around 10 Torr, compared with expected pre-exercise values of approximately 2-4 Torr for carbon dioxide and 100-110 Torr for oxygen (Randall, Perry and Heming 1982). Metabolically, blood lactate increased in a linear fashion with the length of time to capture (Figure 3.1), rising from around 5 μmol ml^{-1} under 'zero activity' condition to over 20 μmol ml^{-1} after 1 h of severe exercise stress. Such high concentrations of blood lactate in fish appear to be related to their increased scopes for activity (see Wells and Davie 1985) compared with less athletically active fish. Unfortunately it is difficult to differentiate between direct metabolic and respiratory contributions to the measured values and those resulting from inadequate gill ventilation prior to blood sampling. When lactic acid is produced, it dissociates into lactate anions and H$^+$ (hydrogen) ions, and there can be differential release rates of H$^+$ and lactate from muscle into the blood (Bilinski, 1974; Wardle, 1978 see also Jensen, Nikinmaa and Weber 1983).

Although marlin exhibit rather extreme acidaemic conditions following capture, the blood does have a very high non-bicarbonate buffering capacity (cf. Table 3.1 as compared with other active teleosts (see Perry, Daxboeck, Emmett, Hochachka and Brill 1983a, b)). This is no doubt due to the high haematocrit and haemoglobin concentration (Wells, Ashby, Duncan and Macdonald 1980), which would remove some of the excess H$^+$ ions from the circulation. None the less, the resultant decreased

Figure 3.1: The Effect of Catch Time on Blood Lactate Concentrations in Blue Marlin (*M. nigricans*); Each Point Representing Duplicate Sample Determinations. Redrawn from Dobson *et al.* (1985)

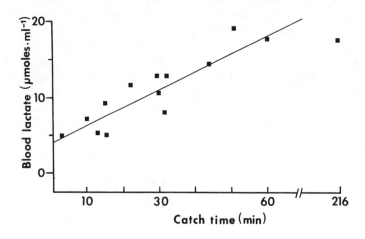

blood pH will still affect the oxygen dissociation curve. Despite relatively high carrying capacity, marlin blood exhibits very marked Root and Bohr effects (cf. Table 3.1 and Wells and Davie 1985), which will affect oxygen uptake and delivery, especially under exercise stress conditions. In addition, capture stress and the attendant acid-base disturbances may cause erythrocytic swelling (Nikinmaa 1981; Wells and Davie 1985), which in turn may be partially responsible for changes in the oxygen dissociation curve.

Length of time of exercise bout is related to a marked decrease in plasma colloid osmotic pressure (Figure 3.2) in marlin (Hargens 1984). However, there are no significant changes in either the blood haematocrit or haemoglobin concentration with increases in time for vigorous exercise. This is despite the fact that the spleen has been shown to significantly decrease exponentially in weight over the same time period (Figure 3.3) (see also Yamamoto, Itazawa and Kobayashi 1980, 1983)

Exercise Performance and Blood Chemistry

By incorporating pre-exercise values for blood pH and P_{aCO_2} from other active teleosts (see Randall *et al.* 1982), the predicted total carbon dioxide for marlin blood should be approximately 8.0 mM. Given an *in vitro* non-bicarbonate blood buffering capacity of 21.3 Slykes, a pure respiratory exercise acidosis needed to depress blood pH to the measured values would correspond to a total blood carbon dioxide of about 20 mM. Since measured total carbon dioxide was much lower than the empirically derived value (Table 3.1), it represents a base deficit and points to blood acidosis of metabolic origin as a major component of the noted low pH. Alternatively, a significant portion of total blood bicarbonate (HCO_3^-) may have been transferred to intracellular or other extracellular compartments, including the environment (Dobson *et al.* 1985). The direct correlation between blood lactate and capture time (Figure 3.1) also supports the idea of a very significant metabolic component to blood acidosis. This relationship demonstrates activation of anaerobic glycogenolysis in muscle during the initial stages of exercise with the concomitant slow release of lactate and fast export of H^+ into the blood as time of exercise increases. The low blood pH noted for marlin captured after only 20 min or less of severe exercise (G.P. Dobson, unpublished data) is consistent with the rapid mobilisation of anaerobic glycolysis, since H^+ ions are known to efflux more rapidly than lactate from white muscle to

Figure 3.2: The Effect of Catch Time on Plasma Colloid Osmotic Pressure (COP) in Blue Marlin (*M. nigricans*; each point representing duplicate determinations for individual fish). Redrawn from Hargens (1984)

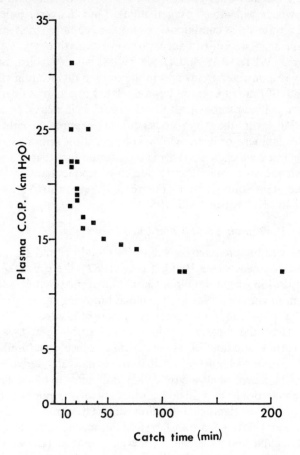

blood in some active fish (Turner and Wood 1983; Turner, Wood and Clark 1983; Wood, Turner and Graham 1983; Perry *et al.* 1985a).

The high buffering capacity of marlin blood compared with other teleosts (see Perry *et al.* 1985a) is an important strategy for extending muscle performance. However, a consequence of scaling up blood buffering capacity is the need to increase pH sensitivity of the haemoglobin for oxygen release, so that its delivery to working muscle tissues will not be compromised. There is no doubt

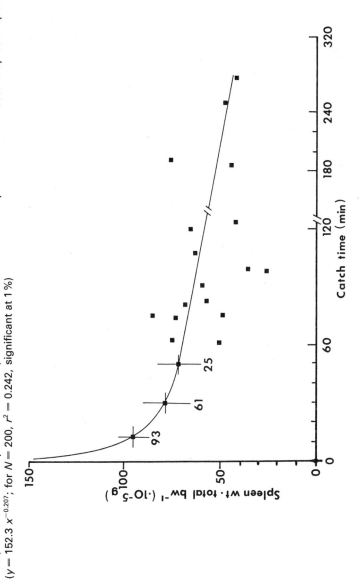

Figure 3.3: The Effect of Catch Time on Spleen Weight in Blue Marlin (*M. nigricans*). Where noted, points are means ± 2 SEM with the number of samples below. Other points are from individual fish. Equation of line for best fit by least squares ($y = 152.3 \, x^{-0.207}$; for $N = 200$, $r^2 = 0.242$, significant at 1 %)

that marlin blood is pH sensitive given the large Bohr and Root effects noted (Dobson *et al.* 1985; Wells and Davie 1985), and these probably operate regardless of whether pH alterations result from fixed acid loads or increased carbon dioxide (Jensen and Weber 1982; Bridges, Hlastala, Riepl and Scheid 1983). The combined Bohr and Root effects in exercise-stressed marlin may not allow adequate blood oxygenation at the gills of marlin. However, the high Bohr coefficient noted for marlin blood would facilitate oxygen delivery and release to aerobic red muscles for a small blood pH change during exercise. The Root effect may have functional significance only for regulation of swim bladder gas composition (Fänge 1976) in marlin. It appears that marlin follow the trend displayed by other teleosts with high scopes for activity, namely low blood oxygen affinities, large Bohr effects and co-operative oxygen binding (Powers 1980; Dobson *et al.* 1985; Wells and Davie 1985). However, data from mako sharks (*Isurus oxyrhinchus*) with a lifestyle similar to that of marlin suggest that, at least for elasmobranchs, a low blood-oxygen affinity and large Bohr effect are not necessarily requirements for an efficient oxygen transport system in fish which also have an extensive scope for activity (Wells and Davie 1985).

It is interesting to note that the oxygen binding capacity of marlin blood appears to be unaffected by temperature change over the range 20-35°C ($n = 21$ fish; R.K. Dupre, S.C. Wood and C. Clark, unpublished data). Although the blood may not be fully saturated under severe exercise conditions, by the reasoning above, the relative temperature insensitivity would prevent even more drastic changes in oxygen uptake at the gills, which are at ambient temperatures, and oxygen delivery to the warmer exercising muscles (2-3°C, Lindsey 1968).

Marlin blood increases in viscosity in an exponential fashion with increasing haematocrit much more rapidly than does mammalian (rat) blood over the same red cell volume range (S.C. Wood, unpublished data). Therefore, after a 'critical' circulating haematocrit is reached, the increased buffering and oxygen transport capacities are outweighed by potential slowing of flow through capillaries, which may lead to local ischaemia, especially in the myocardium via the coronary circulation. Increases in circulating haematocrit in exercising marlin are not supported by available data. On the one hand, decreased plasma colloid osmotic pressure and increased capillary fluid filtration could lead to some local haemoconcentrations, but may not be recorded in the general circulation (Hargens 1984). On the other hand, splenic contraction has been shown to add extra blood to the general circulation during severe exercise in other fish (Yamamoto *et al.* 1980, 1983). The marlin spleen does decrease in weight with increasing exercise duration but given its size compared with the total blood volume ($\approx$5 per cent) of marlin, the ultimate contribution of volume ejection would amount to an insignificant 0.5 per cent increase in the total circulating blood cell volume. The functional significance of splenic contraction to marlin exercise performance therefore remains to be elucidated.

Locomotor Muscles

General Considerations

Most fish swim through water by pushing against this incompressible medium with undulations of their body or fins. These propulsive surfaces produce a forward thrust which is generated by muscular contractions. That marlin accomplish this very well has already been illustrated. In general the scombrids (tunas, swordfish, billfish) swim with stiff bodies (carangiform locomotion), and propulsive motion is restricted to the caudal peduncle and large lunate tail.

The locomotion trunk muscles are organised into segmentally arranged Σ-shaped myotomes. The fibres that constitute the myotomal mass can be characterised by colour and histochemical properties into red (slow, aerobic), pink (fast, aerobic) and white (fast, glycolytic) (see Johnston 1982, 1983). The two most distinctly different muscle types — red and white — have different degrees of vascularisation and myoglobin content, which accounts for their colour.

Force Generation and Power Output

In marlin the locomotor muscle mass comprises between 55 and 75 per cent of body weight and of this between 5 and 7 per cent is 'classical' red muscle (Johnston and Salamonski 1984; Scrimgeour 1984). In this regard the marlins are similar to tunas (Fierstine and Walters 1968; Graham and Diener 1978), except for the fact that tuna 'white' muscle is much more extensively vascularised.

Maximum isometric tensions and power output for marlin red

and white muscle (cf. Table 3.2) are comparable to values obtained from other fish (Altringham and Johnston 1982). Values for maximum contraction velocity (V_{max}) are higher than for most cold-temperate species at temperatures from 0 to 10°C, but are similar to those found in other tropical species at their normal environmental temperature of 25°C (Altringham and Johnston 1982; Johnston and Brill 1984; Johnston and Altringham 1985). Maximum mechanical power output of rainbow trout white muscle has been estimated at 50 Wkg^{-1} body weight at 12°C (see Johnston and Salamonski 1984 for details), a figure comparable to that obtained for billfish (Table 3.2), but much higher than for the white muscle of non-athletic fishes. Additionally, maximum power output for marlin white muscle is around four times higher than for red fibres. The power output and tensions generated for both fibre types, however, are independent of temperature over a range of 15-30°C (Johnston and Salamonski 1984; Altringham and Johnston 1985).

Assuming muscle comprises 75 per cent of body mass in marlin and 93 per cent of this is white muscle, then the total mechanical power output available is about 2.8 kW for a 70 kg fish, rising to 20 kW for a 455 kg (1000 lb) specimen (equivalent to 27 h.p.; Johnston and Salamonski 1984).

Energy Metabolism and Muscle Buffering Capacities

Enzyme activities in marlin red and white muscles (see page 65) are similar to those reported for tuna (Guppy, Hulbert and Hochachka 1979). High glycolytic capacity in white muscle is indicated by high pyruvate kinase (PK) and lactate dehydrogenase (LDH) activities. In addition, white muscle also has a substantial capacity for aerobic energy production by carbohydrate (glucose and/or glycogen) oxidation. Fats and amino acids are relatively unimportant substrates for this tissue, unlike marlin red muscle (Suarez *et al.* 1985).

Phosphocreatine concentrations in teleost white muscles are typically around 20 mmol kg^{-1} (Walesby and Johnston 1980). This would be sufficient for 14.5 s of maximal power output (Johnston and Salamonski 1984). Perhaps the acrobatic tail-walking sometimes exhibited by marlin is performed at less than maximum to conserve fuel reserves for longer sprints. Half maximal power output would be equivalent to a glycolytic flux of 0.75 mmol glucose kg^{-1}s^{-1} but a 30-min exercise bout results in measured

Table 3.2: Contractile Properties of Skinned Fibres Isolated from the Trunk Muscles of Blue (*M. nigricans*) and Striped (*T. audax*) Marlin. (Data from: Johnston and Salamonski (1984 and unpublished); Johnston and Altringham (1985); Altringham and Johnston (1985).)

Parameter	Pacific blue marlin (25°C)		Pacific blue marlin (15°C)		Striped marlin (25°C)	
	Red fibres	White fibres	Red fibres	White fibres	Red fibres	White fibres
Max. Ca^{2+}-activated force (P_0, Ncm^{-2})	5.7 ± 0.9 (11)	17.6 ± 2.1 (13)	5.5 ± 0.6 (11)	15.3 ± 1.0 (10)	5.6 ± 0.7 (5)	17.4 ± 1.1 (5)
V_{max} (L_0s^{-1})*	2.5 ± 0.3 (11)	5.3 ± 0.4 (10)	1.8 (9)	4.0 (10)	3.2 ± 0.6 (5)	7.5 ± 0.7 (5)
Load for maximum power output	0.31 P_0	0.25 P_0	0.31 P_0	0.25 P_0	0.31 P_0	0.25 P_0
Maximum power output (Wkg^{-1})	13.1	57.2	9.7	37.2	15.9	89.9

*L_0 = relaxed muscle fibre length.
Numbers are means ± SEM; those in parenthesis represent number of fish sampled.

blood lactate concentrations of only 8-12 mM, instead of a derived estimate of 4.5 M lactate (I.A. Johnston unpublished data)! Either the fish cannot work at maximum power for very long, or there is an enormous lactate pool in muscle.

Marlin red and white muscles have very high non-bicarbonate buffering capacities (Table 3.3), and white muscle in particular has an anaerobic potential similar to that of tunas (Castellini and Somero 1981). White muscle has a two-fold higher buffering capacity than does red muscle, a finding consistent with other fish species (Abe, Dobson, Hoeger and Parkhouse 1985). This difference becomes significant when considering rates of ATP production in each tissue and the fate of protons generated under working conditions. Marlin white muscle is predominantly fast-twitch glycolytic, which generates a power output of about 50 μmol ATP g^{-1} muscle min⁻ at 25°C. This represents a value about five times that for potential anaerobic red muscle power output (Johnston and Salamonski 1984). A high buffering capacity is therefore related to the high glycolytic potential (see Table 3.4). There are, however, two sources of H^+ ions in working muscle: those generated from anaerobic glycolysis and those generated via aerobic carbon dioxide hydration. White muscle generally has few mitochondria and tissue capillaries compared with red muscle (Johnston 1981a, b), and protons are therefore generated by the coupling of ATP hydrolysis with glycogenolysis (Hochachka and Mommsen 1983). Protons generated during anaerobic glycolysis or fermentation do not arise from dissociation of lactic acid since the end-product is the lactate anion, not free acid (see Gevers 1977). Therefore during burst work the number of protons generated per mole of substrate fermented in white muscle is trivial

Table 3.3: Buffering Capacities of Red and White Skeletal Muscle of the Pacific Blue Marlin (*Makaira nigricans*) and Rainbow Trout (*Salmo gairdneri*). (From Dobson *et al.* (1985).)

Species	Muscle type	Buffering capacity*
Marlin	White	103.8 ± 5.12 (5)
	Red	50.8 ± 3.9 (5)
Trout	White	59.7 ± 0.8 (5)
	Red	31.6 ± 1.1 (5)

*Buffering capacity (β) is expressed in units of μmol of base (NaOH) required to titrate the pH of 1 g of muscle (wet weight) by one pH unit at 25°C. Values given are mean ± SE with the number of fish sampled in parentheses

Table 3.4: Selected Enzyme Activities in Tissues from Pacific Blue Marlin, *M. nigricans*, at 25°C (Data from Suarez *et al.* 1985) Compared with Data from skipjack tuna (from Hochachka *et al.* 1978) in parentheses

Enzyme	White muscle	Activity ($\mu mol\,min^{-1}\,g^{-1} \pm SEM$) ($n = 4$) Red muscle	Heart muscle
Pyruvate kinase (PK)	1105.1 ± 64 (1294.9)	315.9 ± 37 (195.2)	103.9 ± 12 (126.6)
Lactate dehydrogenase (LDH)	2696.6 ± 125 (5492.3)	616.6 ± 93 (514.4)	707.8 ± 36 (449.0)
Phosphoenolpyruvate carboxykinase (PEPCK)	0.90 ± 0.28	not detectable	not detectable
Fructose 1, 6-biphosphatase (FBPase)	2.39 ± 0.55	0.68 ± 0.19	not detectable
Creatine phosphokinase (CPK)	837.2 ± 40 (516.4)	1433.7 ± 192 (554.2)	172.5 ± 24 (115.2)

(Hochachka and Mommsen 1983) compared with the amount generated by coupling ATP hydrolysis. Once white-muscle buffering capacity is saturated, intra muscular pH falls and H^+ ions are forced down their concentration gradient into the circulation.

In red muscle under strictly aerobic conditions there is no net H^+ production since release rate of ATP hydrolysis is matched by H^+ consumption during mitochondrial oxidative phosphorylation (Krebs, Woods and Alberti 1975). Therefore the major source of protons during high aerobic work rates in red muscle is from hydration of metabolically produced carbon dioxide. The buffering capacity therefore need not be as high as in white muscle. However, marlin red muscle still exhibits higher buffering capacity than red muscle of trout (Table 3.3). In the marlin, then, the red muscle is geared for both aerobic and anaerobic ATP generation at high muscle work rates (see also Table 3.4: enzyme profiles, page 65).

Energy Metabolism and Endothermy

Muscles are generally less than 25 per cent efficient and a large part of the energy from chemical breakdown goes into heat production rather than mechanical work (Johnston 1982). Endothermy requires heat production and temperature regulation. Tuna red and white muscles have unusually high capacities for aerobic metabolism sufficient to account for elevated temperatures (Hochachka 1974; Guppy *et al.* 1979). There is great similarity between the enzymatic make-up, therefore metabolic potential, of tuna and marlin (cf. Table 3.4), and the potential for metabolic heat production is present in marlin muscle as well. This potential remains unrealised in marlin since they cannot retain muscle heat. Marlin have no countercurrent heat exchangers or retes in red muscles as are found in tunas (Carey 1982a), and fish lacking muscle retes lose metabolic heat through the body wall, fins and gills (Stevens and Sutterlin 1976). Marlin do, however, have thermogenic tissue and vascular retes to conserve the heat produced at the back of the eye near the brain (Block 1983, 1984; Carey 1982b).

It would appear that biochemical adaptations for a pelagic, fast-swimming lifestyle are similar to those required for thermogenesis. Tunas exploit this thermogenic potential by conserving metabolic muscle heat through retes, whereas marlin have cold muscle and only keep their eyes and brains above ambient. It is interesting to note that contraction of tuna muscle is more temperature sensitive than in other fishes (Johnston and Brill 1984) while in marlin it is less sensitive to temperature within their environmentally encountered temperature range (Johnston and Salamonski 1984).

Hearts

General Considerations

The ventricle is an elongated pyramidal shape. It comprises at least 60 per cent of the total heart weight and in marlin represents approximately 0.15 per cent of the total body mass (Davie and Daxboeck 1984; Mason 1984). Its shape and high relative weight are typical of very active teleosts, and place it with hearts of tunas and small mammals (Poupa, Lindström, Maresca and Tota 1981; Poupa and Lindström 1983). The ventricle wall has an outer shell of compact muscle which comprises 27-35 per cent by weight of the total ventricular mass in marlin (Mason 1984) and is high for fish (see Tota 1983). The outer shell overlies the inner spongy myocardium, composed of muscular trabeculae. The spaces or sinusoids between trabeculae are filled with the venous blood of the heart lumen.

Ventricular volumes of marlin, and hence maximum stroke volumes, are estimated at around $0.23\,ml\,kg^{-1}$ body weight (Mason 1984). Rainbow trout have a ventricular volume of approximately $1.0\,ml\,kg^{-1}$ body weight (Wood and Shelton 1980; P.S. Davie, unpublished data). The anatomical evidence suggests that the marlin heart is a high-pressure, high-stroke-rate pump. However, no *in vivo* measurements are available to confirm this conclusion.

Coronary Blood Supply

The coronary arterial supply in marlin has both cranial and caudal sources (Mason 1984). In most fishes the coronary blood supply has only a cranial origin (Watson and Cobb 1979). The caudal coronary arteries arise from the first branch of the dorsal aorta (Mason 1984). The cranial coronary supply is derived primarily from the ventral branches of the second efferent branchial arteries with minor contributions from a common trunk of the third and fourth efferent branchial arteries. Several branches pass superficially over the pericardium and anastomose with the caudal coronary supply, forming an elaborate vascular net around the myocardium. These arteries branch to invest all layers of the ventricle with an extensive arterial network. Smaller arteries have been shown to terminate on myocardial sinusoids between trabeculae (Mason 1984), and smaller arterioles anastomose with terminal branches of the sinusoids. This arrangement completes a Thesbesian vascular system in the marlin heart (Davie and Daxboeck 1984), as also reported for swordfish (*Xiphius gladius*) by Tota (1983).

Coronary Vascular Activity

There is a paucity of information concerning the function and regulation of coronary blood flow in fishes. The literature contains references to reactivity of coronary vasculature to vasoactive substances in the Pacific blue marlin (Davie and Daxboeck 1984), the Atlantic salmon (Farrell and Graham 1985) and the conger eel (Belaud and Peyraud 1971). In addition, Daxboeck (1982) has discussed the possible function of the coronary blood supply during exercise in rainbow trout.

The marlin coronary vascular bed has an adrenergic pharmacology similar to that of higher vertebrates, despite the 'remote branchial origin' of its blood supply (Grant and Regnier 1926). These beds contain β-dilator and α-constrictor adrenoceptors (Figure 3.4), and, as in the mammalian case, adrenaline is a more potent α-agonist than is noradrenaline (Zuberbuhler and Bohr 1965). Experiments by Davie and Daxboeck (1984) do not distinguish between β_1- or β_2-dilator receptors, and further investigation is needed to elucidate this point. Although *in vivo* levels of circulating catecholamines are unavailable for marlin, the threshold and half maximal doses of adrenaline and noradrenaline for coronary vasoactivity are similar to other teleost vascular preparations (Davie 1981; Canty and Farrell 1985; Farrell and Graham 1985), and to mammalian coronary preparations (Feigl 1983).

How coronary blood supply to the two myocardial layers is controlled is an extremely interesting question. Circulating catecholamines or sympathetic nerve stimulation may produce a moderate coronary vasoconstriction since adrenaline is the dominant nerve-terminal catecholamine in fish hearts (Ask 1983). However, this may be overridden by concomitant positive inotropic and chronotropic effects of catecholamines on the myocardium, which increase heart metabolism and may lead to local active hyperaemic vasodilation (Farrell and Graham 1985), as suggested by Randall

Figure 3.4: Cumulative Dose-Response Curves of Vascular Resistance Changes in Perfused Marlin (*M. nigricans*) Coronary Vessels in Response to Adrenaline (AD, *n* = 5), Noradrenaline (NA, *n* = 5), and Isoprenaline (ISO, *n* = 4). Points are means ± 1 SEM. Redrawn from Davie and Daxboeck (1984)

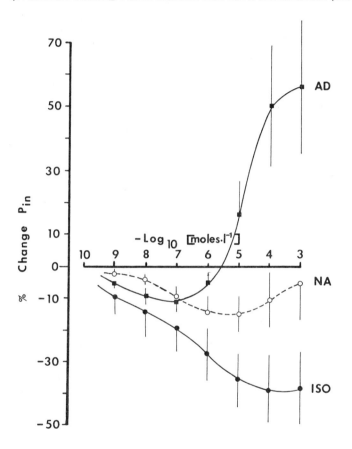

and Daxboeck (1982) for rainbow trout trunk musculature during exercise.

Myocardial Energy Metabolism

The marlin myocardium, like marlin red muscle, is a highly aerobic tissue which continuously performs work. The primary energy for myocardial activity is derived from aerobic metabolism of fats, amino acids and carbohydrates (Suarez *et al.*, 1985). Both phosphoenolpyruvate carboxykinase (PEPCK) and fructose

1,6-biphosphatase (FBPase) are absent and PK activities are low in marlin myocardium (see Table 3.4). It would appear then that, should heart muscle glycogen stores become depleted, perhaps after a burst of activity, they must be restored aerobically using glucose from the blood. Unlike white muscle, which has high levels of PK activity, the myocardial PK levels are low. A reverse flux of the PK-activated reaction for lactate conversion to glycogen may occur at 1 per cent of the forward reaction (Dyson, Cardenas and Barsotti 1975), but since myocardial PK levels are minimal, there would be no recovery of glycogen stores by this anaerobic pathway in the marlin heart. In addition, myocardial creatine phosphokinase (CPK) activities are very low and therefore involvement of a creatine phosphate shuttle (Bessman and Geiger 1981) in the energy metabolism of this muscle tissue is unlikely.

The marlin myocardium also has significant LDH activities (Table 3.4). Although it is not known whether the myocardial LDH is the same kind as found in red or white muscles (see Sidell 1983), its presence in the heart muscle does indicate some degree of anaerobic glycolytic capacity. Unfortunately, the study of Suarez *et al.* (1985) does not differentiate enzyme activities between spongy and compact myocardium. Fortunately Tota (1978, 1983) and Hochachka, Hulbert and Guppy (1978) have presented such data for tunas: ecologically and physiologically similar teleosts. These authors concluded that the spongy myocardium would be exposed to a microenvironment of low P_{O_2}, lower pH, accumulation of metabolites and lactate, etc. in active fish. Therefore enzyme systems designed to oxidise lactate (LDH) more efficiently under more anaerobic conditions would be of obvious advantage. It has been noted that, following periods of high activity, the venous blood oxygen content of marlin approaches zero (S.F. Perry, unpublished data). Hence the spongy myocardium would be working under almost anoxic conditions and the aforementioned LDH activity would certainly come into play. However, it has been stated previously that the marlin heart has an extensive coronary arterial vasculature which extends into the trabecular myocardium. This additional blood supply and possible redistribution of regional coronary blood flow from the compact layer to favour the inner spongy layer under extreme exercise situations may allow maintenance of aerobic metabolism of both layers under such conditions.

References

Abe, H., Dobson, G.P., Hoeger, H. and Parkhouse, W.S. (1985) 'The Role of Histidine and Histidine Related Compounds to Intra-cellular Buffering Fish Skeletal Muscle', *American Journal of Physiology*, in press

Altringham, J.D. and Johnston, I.A. (1982) 'The pCa-Tension and Force-Velocity Characteristics of Skinned Fibres Isolated from Fish Fast and Slow Muscles', *Journal of Physiology (London), 333*, 421-49

Altringham, J.D. and Johnston, I.A. (1985) 'The Effects of Temperature on ATPase Activity and Force Generation in Skinned Muscle Fibres from the Pacific Blue Marlin (*Makaira nigricans*)', *Experimentia*, in press

Ask, J.A. (1983) 'Comparative Aspects of Adrenergic Receptors in the Hearts of Lower Vertebrates', *Comparative Biochemistry and Physiology, 76A*, 543-52

Belaud, A. and Peyraud, C. (1971) 'Etude preliminaire du debit corenair sur coeur perfuse de Poisson', *Journal of Physiology (Paris), 63*, 165A

Bessman, S.P. and Geiger, P.J. (1981) 'Transport of Energy in Muscle: the Phosphorylcreatine Shuttle', *Science, 211*, 448-52

Bilinski, E. (1974) 'Biochemical Aspects of Fish Swimming', in D.C. Malins and J.R. Sargent (eds), *Biochemical and Biophysical Perspectives in Marine Biology*, Academic Press, New York

Block, B.A. (1983) 'Brain Heaters in Billfish', *American Zoologist, 23* (4), 936

Block, B.A. (1984) 'A Thermogenic Tissue in Billfish: Do Fish Have Brown Fat?', *American Zoologist, 24* (3), 98A

Bridges, G.R., Hlastala, M.P., Riepl, G. and Scheid, P. (1983) 'Root Effect Induced by CO_2 and by Fixed Acid in the Blood of the Eel. *Anguilla anguilla*', *Respiration Physiology, 51*, 275-86

Canty, A.A. and Farrell, A.P. (1985) 'Intrinsic Regulation of Flow in an Isolated Tail Preparation of the Ocean Pout (*Macrozoarces americanus*)', *Canadian Journal of Zoology, 63*, in press

Carey, F.G. (1982a) 'Warm Fish', in C.R. Taylor, K. Johansen and L. Bolis (eds), *A Companion to Animal Physiology*, Cambridge University Press, Cambridge

Carey, F.G. (1982b) 'A Brain Heater in the Swordfish', *Science, 216*, 1327-9

Castellini, M.A. and Somero, G.N. (1981) 'Buffering Capacity of Vertebrate Muscle: Correlations with Potentials for Anaerobic Function', *Journal of Comparative Physiology, 143*, 191-8

Davie, P.S. (1981) 'Vascular Resistance Responses of an Eel Tail Preparation: Alpha Constriction and Beta Dilation', *Journal of Experimental Biology, 90*, 65-84

Davie, P.S. and Daxboeck, C. (1984) 'Anatomy and Adrenergic Pharmacology of the Coronary Vascular Bed of Pacific Blue Marlin (*Makaira nigricans*)', *Canadian Journal of Zoology, 62*, 1886-8

Daxboeck, C. (1982) 'The Effect of Coronary Artery Ablation on Exercise Performance in the Rainbow Trout', *Canadian Journal of Zoology, 60*, 375-81

Dobson, G.P., Wood, S.C., Daxboeck, C. and Perry, S.F. (1985) 'Intracellular Buffering and Oxygen Transport in the Pacific Blue Marlin (*Makaira nigricans*): Adaptations to High-speed Swimming', *Physiological Zoology*, in press

Dyson, R.D., Cardenas, J.M. and Barsotti, R.J. (1975) 'The Reversibility of Muscle Pyruvate Kinase and an Assessment of its Capacity to Support Glycogenesis', *Journal of Biological Chemistry, 250*, 3316-21

Fänge, R. (1976) 'Physiology of the Swimbladder', *Physiological Review, 46*, 299-322

Farrell, A.P. and Graham, M.S. (1985) 'Effects of Adrenergic Drugs on the

Coronary Circulation of Atlantic Salmon (*Salmo salar*)', *Canadian Journal of Zoology*, in press

Feigl, E.O. (1983) 'Coronary Physiology', *Physiological Review*, *63*, 1-205

Fierstine, H.L. and Walters, V. (1968) 'Studies in Locomotion and Anatomy of Scombrid Fish', *Memoirs of the Southern California Academy of Science*, *6*, 1-31

Gevers, W. (1977) 'Generation of Protons by Metabolic Processes in Heart Cells', *Journal of Molecular and Cellular Cardiology*, *9*, 867-74

Graham, M.S. and Diener, D.R. (1978) 'Comparative Morphology of the Central Heat Exchangers in the Skipjacks *Katsuwonus* and *Euthynnus*', in G.D. Sharp and A.E. Dizon (eds), *The Physiological Ecology of Tunas*, Academic Press, New York

Grant, R.T. and Regnier, M. (1926) 'The Comparative Anatomy of the Cardiac Vessels', *Heart*, *13*, 285-317

Guppy, M., Hulbert, W.C. and Hochachka, P.W. (1979) 'Metabolic Sources of Heat and Power in Tuna Muscles II. Enzyme and Metabolite Profiles', *Journal of Experimental Biology*, *82*, 303-20

Hargens, A.R. (1984) 'Decreased Colloid Osmotic Pressure in Blood of Pacific Marlin *Makaira nigricans* Related to Duration of Fighting Exercise', *Medical Science in Sports and Exercise*, *16* (2), 138-9

Hochachka, P.W. (1974) 'Regulation of Heat Production at the Cellular Level', *Federation Proceedings*, *33*, 2162-9

Hochachka, P.W. and Mommsen, T.P. (1983) 'Protons and Anaerobiosis', *Science*, *219*, 1391-7

Hochachka, P.W., Hulbert, W.C. and Guppy, M. (1978) 'The Tuna Power Plant and Furnace', in G.D. Sharp and A.E. Dizon (eds), *The Physiological Ecology of Tunas*, Academic Press, New York

Jensen, F.B. and Weber, R.E. (1982) 'Respiratory Properties of Tench Blood and Hemoglobin: Adaptation to Hypoxic-Hypercapnic Water', *Molecular Physiology*, *2*, 235-50

Jensen, F.B., Nikinmaa, M. and Weber, R.E. (1983) 'Effects of Exercise Stress on Acid-Base Balance and Respiratory Functions in Blood of the Teleost *Tinca tinca*', *Respiratory Physiology*, *51*, 291-301

Johnston, I.A. (1981a) 'Structure and Function of Fish Muscles', in M.H. Day (ed.), *Symposium of the Zoological Society of London. Vertebrate Locomotion*, *48*, 71-113

Johnston, I.A. (1981b) 'Specialization of Fish Muscle', in D.F. Goldspink (ed.), *Development and Specialization of Muscles*, Cambridge University Press, Cambridge

Johnston, I.A. (1982) 'Biochemistry of Myosins and Contractile Properties of Fish Skeletal Muscle', *Molecular Physiology*, *2*, 15-29

Johnston, I.A. (1983) 'On the Design of Fish Myotomal Muscle', *Marine Behavioural Physiology*, *9*, 83-98

Johnston, I.A. and Altringham, J.D. (1985) 'Evolutionary Adaptation of Muscle Power Output to Environmental Temperature: Force-Velocity Characteristics of Skinned Fibres Isolated from Antarctic, Temperate and Tropical Marine Fish', *Pflügers Archiv*, in press

Johnston, I.A. and Brill, R.W. (1984) 'Thermal Dependence of Contractile Properties of Single-skinned Muscle Fibres Isolated from Antarctic and Various Warm-water Marine Fishes Including Skipjack Tuna (*Katsuwonus pelamis*) and Kawakawa (*Euthynnus affinis*)', *Journal of Comparative Physiology*, *155*, 63-70

Johnston, I.A. and Salamonski, J. (1984) 'Power Output and Force-Velocity Relationship of Red and White Muscle Fibres from the Pacific Blue Marlin

(*Makaira nigricans*)', *Journal of Experimental Biology, 111*, 171-7

Jones, D.R. and Randall, D.J. (1978) 'The Respiratory and Circulatory Systems during Exercise', in W.S. Hoar and D.J. Randall (eds), *Fish Physiology*, vol. 7, Academic Press, New York

Krebs, H.A., Woods, H.F. and Alberti, K.G.M.M. (1975) 'Hyperlactataemia and Lactic Acidosis', *Essays in Medical Biochemistry, 1*, 81-103

Lindsey, C.C. (1968) 'Temperatures of Red and White Muscle in Recently Caught Marlin and Other Large Tropical Fish', *Journal of the Fisheries Research Board of Canada, 25* (9), 1987-92

Magnuson, J.J. (1978) 'Locomotion by Scombrid Fishes: Hydromechanics, Morphology, and Behavior', in W.S. Hoar and D.J. Randall (eds), *Fish Physiology*, vol. 7, Academic Press, New York

Mason, R.H. (1984) 'Comparative Heart Anatomy of the Striped Marlin' (*Tetrapterus audax*) and the Mako Shark (*Isurus oxyrhinchus*) with Special Reference to the Coronary Circulation', B. Phil. thesis, Massey University, New Zealand

Mather, C.O. (1976) '*Billfish*', Saltaire Publishing, Sidney BC

Nikinmaa, M. (1981) 'Respiratory Adjustments of Rainbow Trout (*Salmo gairdneri* Richardson) to Changes in Environmental Temperature and Oxygen Availability' Ph. D. thesis, University of Helsinki, Finland

Perry, S.F., Daxboeck, C., Emmett, B., Hochachka, P.W. and Brill, R.W. (1985a) 'Effects of Exhausting Exercise on Acid-Base Regulation in Skipjack Tuna (*Katsuwonus pelamis*) Blood', *Physiological Zoology, 58*, (4), 421-9

Perry, S.F., Daxboeck, C., Emmett, B., Hochachka, P.W. and Brill, R.W. (1985b) 'Effects of Temperature Change on Acid-Base Regulation in Skipjack Tuna (*Katsuwonus pelamis*) Blood', *Comparative Biochemistry and Physiology, 81A*, 49-53

Poupa, O. and Lindström, L. (1983) 'Comparative and Scaling Aspects of Heart and Body Weights with Reference to Blood Supply of Cardiac Fibres', *Comparative Biochemistry and Physiology, 76A*, 413-21

Poupa, O., Lindström, L., Maresca, A. and Tota, B. (1981) 'Cardiac Growth, Myoglobin, Proteins and DNA in Developing Tuna (*Thunnus thynnus thynnus* L.)', *Comparative Biochemistry and Physiology, 70A*, 217-22

Powers, D.D. (1980) 'Molecular Ecology of Teleost Fish Hemoglobins: Strategies for Adapting to Changing Environments', *American Zoologist, 20*, 139-62

Randall, D.J. and Daxboeck, C. (1982) 'Cardiovascular Changes in the Rainbow Trout (*Salmo gairdneri* Richardson) during Exercise', *Canadian Journal of Zoology, 60*; 1135-40

Randall, D.J., Perry, S.F. and Heming, T.A. (1982) 'Gas Transfer and Acid-Base Regulation in Salmonids', *Comparative Biochemistry and Physiology, 73B*, 93-103

Scrimgeour, A.B. (1984) 'Aspects of the Anatomy of the Striped Marlin (*Tetrapterus audax*. Phillipi, 1887) and the Black Marlin (*Makaira indica* Cuvier, 1831)', B. Phil. thesis, Massey University, New Zealand

Sidell, B.D. (1983) 'Cardiac Metabolism in the Myxinidae: Physiological and Phylogenetic Considerations', *Comparative Biochemistry and Physiology 76A*, 495-505

Stevens, E.D. and Sutterlin, A.M. (1976) 'Heat Transfer between Fish and Ambient Water', *Journal of Experimental Biology, 65*, 131-45

Suarez, R.K., Mallet, M.D., Daxboeck,C. and Hochachka, P.W. (1985) 'Enzymes of Energy Metabolism and Gluconeogenesis in the Pacific Blue Marlin *Makaira nigricans*', (manuscript submitted)

Tota, B. (1978) 'Functional Cardiac Morphology and Biochemistry in Atlantic Bluefin Tuna', in G.D. Sharp and A.E. Dizon (eds), *The Physiological Ecology* of Tunas, Academic Press, New York

Tota, B. (1983) 'Vascular and Metabolic Zonation in the Ventricular Myocardium of Mammals and Fishes', *Comparative Biochemistry and Physiology, 76A*, 423-37

Turner, J.D. and Wood, C.M. (1983) 'Factors Affecting Lactate and Proton Efflux from Pre-exercised, Isolated, Perfused Rainbow Trout Trunks,' *Journal of Experimental Biology, 105*, 395-401

Turner, J.D., Wood, C.M. and Clark, D. (1983) 'Lactate and Proton Dynamics in the Rainbow Trout (*Salmo gairdneri*)', *Journal of Experimental Biology, 104*, 247-68

Walesby, N.J. and Johnston, I.A. (1980) 'Temperature Acclimation on Brook Trout Muscle. Adenine Nucleotide Concentrations, Phosphorylation State and Adenylate Energy Change', *Journal of Comparative Physiology, 139*, 127-33

Wardle, C.S. (1978) 'Non-release of Lactic Acid from Anaerobic Swimming Muscle of Plaice *Pleuronectes platessa* L.: a Stress Reaction', *Journal of Experimental Biology, 77*, 141-55

Watson, A.D. and Cobb, J.L.S. (1979) 'A Comparative Study of the Innervation and Vascularization of the Bulbus Arteriosus in Teleost Fish', *Cell Tissue Research, 196*, 337-46

Wells, R.M.G. and Davie, P.S. (1985) 'Oxygen Binding by the Blood and Hematological Effects of Capture Stress in Two Big Gamefish: Mako Shark and Striped Marlin', *Comparative Biochemistry and Physiology, 81A*, 643-6

Wells, R.M.G., Ashby, M.D., Duncan, S.J. and Macdonald, J.A. (1980) 'Comparative Study of the Erythrocytes and Haemoglobins in Nototheniid Fishes from Antarctica', *Journal of Fish Biology, 17*, 517-27

Wood, C.M. and Shelton, G. (1980) 'Cardiovascular Dynamics and Adrenergic Responses of the Rainbow Trout *in vivo*', *Journal of Experimental Biology, 87*, 247-70

Wood, C.M., Turner, J.D. and Graham, M.S. (1983) 'Why Do Fish Die after Severe Exercise?', *Journal of Fish Biology, 22*, 189-201

Yamamoto, K.I., Itazawa, Y. and Kobayashi, H. (1980) 'Supply of Erythrocytes into the Circulating Blood from the Spleen of Exercised Fish', *Comparative Biochemistry and Physiology, 65A*, 5-11

Yamamoto, K.I., Itazawa, Y. and Kobayashi, H. (1983) 'Erythrocyte Supply from the Spleen and Hemoconcentration in Hypoxic Yellowtail', *Marine Biology, 73*, 221-6

Zuberbuhler, R.C. and Bohr, D.F. (1965) 'Responses of Coronary Smooth Muscle to Catecholamines', *Circulation Research, 16*, 431-40

4 FISH CARDIOLOGY: STRUCTURAL, HAEMODYNAMIC, ELECTROMECHANICAL AND METABOLIC ASPECTS

Kjell Johansen and Hans Gesser

This chapter is an attempt to focus on recent advances in some specific aspects of fish cardiology. One section deals with haemo-dynamic characteristics of the fish heart and the role played in this respect by the pericardium and the four contractile segments of the heart: the sinus venosus, the atrium, the ventricle and the bulbus arteriosus (bulbus cordis in elasmobranchs). The analysis will be restricted to teleost and elasmobranch fishes.

Another section will mainly concern electromechanical coupling and cellular energy liberation under 'normal' conditions and at different loads on the myocardium. Sections addressing some cellular aspects of myocardial function in teleosts are included in the chapter. The analysis is based mainly on mechanical properties of cardiac muscle strips studied *in vitro*, and biochemical studies of the fish myocardium.

Finally some metabolic aspects of myocardial function will be discussed in a comparative context.

Structural Design of the Heart

The fish heart typically consists of four chambers coupled in series. The sinus venosus, sparsely equipped with cardiac muscle, receives central venous blood from the greater veins through the paired ducts of Cuvier. The muscular and often diverticulated atrium has a luminal size similar to the ventricle, and receives blood via the sinus venosus through an ostium guarded by flap valves (sinu-artrial valves). The atrium communicates with the ventricle through the atrioventricular valves. The competence of both valve sets is aided by sphincteric muscles surrounding them. The ventricle typically in both elasmobranchs and teleosts consists of an outside dense compact layer of myocardial tissue, while a spongy,

71

trabeculate myocardium fills most of the ventricular lumen compartmentalising the blood into small lacunar spaces inside the spongy myocardium. In teleosts blood propagates from the ventricle through semilunar ventriculo-bulbar valves into the highly elastic fourth chamber, the bulbus arteriosus, typically connected to the ventral aorta without valvular support for prevention of backflow during ventricular relaxation. In elasmobranchs the bulbar segment is referred to as the bulbus cordis (conus arteriosus) because its wall is distinctly made up of myocardial tissue in addition to elastic tissue. The elasmobranch bulbus cordis may contain a variable number of valves arranged in tiers. The contraction of this myocardial segment is viewed to serve two functions: one securing the patency of the bulbar valves when the intrapericardial pressure (IPP) is at its lowest, thus preventing backflow to the bulbar area. Secondly, contraction of the bulbus cordis will serve as an auxiliary ejection pump in addition to prolonging flow into the ventral aorta. This is a result of the elastic Windkessel effect of the bulbus (Satchell 1971). In teleosts there are no bulbar valves. The conspicuous rebound of its great elasticity must at physiological heart rates prolong cardiac outflow throughout ventricular diastole, thus preventing reversed flow in the ventral aorta and the possible loss of efficiency in branchial gas exchange.

In the typical fish heart all four chambers play an active role in either the filling of other cardiac chambers or in the ejection of the cardiac stroke volume. In elasmobranchs the sinus venosus may play a lesser role as a contractile chamber and filling agent for the atrium than is the case in teleosts.

All the cardiac chambers of the fish heart are enclosed in a pericardium of fibrous tissue variably adhering to surrounding tissues, making a rigid space around the heart (more rigid in elasmobranchs than teleosts). The pericardial space is filled with fluid,an ultrafiltrate from plasma.

The Cardiac Cycle

The cardiac cycle and the associated haemodynamic events in fishes show specialised features in comparison with other vertebrates, due primarily to the serial arrangement of all four cardiac chambers. Also the short segment of ventral aorta connecting the heart in series with the delicate respiratory exchange surfaces in the gills is unique to fishes. The presence of a rigid or semirigid pericardium and the sinus venosus as a distinct contractile cardiac compartment and the sinu-atrial valves are important determinants of venous return as well as atrial and ventricular filling.

The high resistance of the pericardium to deformation implies that blood ejected from the heart will reduce intrapericardial pressure (IPP), while blood entering the heart from central veins will increase heart size and increase IPP. In elasmobranch fishes a pulsatile subambient IPP persists throughout the cardiac cycle (Johansen 1965a,b; Sudak 1965a,b; Satchell 1971; Shabetai, Abel, Graham, Bhargava, Keyes and Witztum 1985). When IPP remains subambient, it must imply that at normal heart rate blood leaves the heart at a higher volume rate (time course of ejection more rapid than filling) than it enters. Importantly, however, the periods of ejection from the heart overlap with cardiac filling (Satchell 1971). In the teleost fish on which much less work has been done on pericardial function, a subambient IPP prevails during 50 per cent of the cardiac cycle length (Chow and Chan 1975).

A cardiac cycle begins with filling of the sinus venosus from the pressure head in the greater central veins. In these a positive pressure head always prevails towards the sinus venosus except when the latter contracts (teleosts).

When IPP is below ambient pressure, this will affect the transmural pressure as well as the dynamic pressures inside all cardiac chambers enclosed within the pericardial space. The influence of IPP on intracardiac pressures is, as expected, greater in the thin-walled sinus venosus and atrium and to some extent the bulbus compared with the thick-walled ventricle. When a subambient IPP steepens dynamic pressure gradients and influences venous filling, an aspiratory or suctional attraction for blood is in operation (a *vis-à-fronte* movement of blood).

The fluctuations of IPP during the cardiac cycle are largest during ventricular ejection and to some extent during bulbar rebound or contraction, which further reduces the intracardiac blood volume and heart size. During sinus venosus and atrial contraction prior to ventricular contraction and ejection, changes in total intracardiac blood volume and thus IPP should not occur. When this has been demonstrated to occur, however, (Sudak 1965a,b; Chow and Chan 1975; Shabetai *et al.* 1985), the patency of the sinu-atrial and atrioventricular valves are surrounded by sphincteric muscles aiding the passive flap valves which otherwise may be less patent due to the fluctuations in transmural pressures set up by variations in IPP. Other specialisations in valve functions occur on the hepatic veins entering the sinus venosus of elasmo-branchs. These openings have no passive valves but distinct sphincteric muscles controlled cholinergically (Johansen and Hanson 1967), probably serving to adjust hepatic blood flow and functions of the liver.

The end-diastolic volume and associated stretch of the myocardial fibres is, according to the Starling principle, a main determinant of myocardial energy release and hence stroke volume. This principle also applies to the fish heart. In fishes the filling mechanism of the heart, terminating with each cardiac cycle at the ventricular end-diastolic volume, hence becomes decisive in regulation of the stroke volume. The filling of the fish heart shows features distinct from those of other vertebrates, particularly from birds and mammals lacking a sinus venosus and thus sinu-atrial valves. The presence of a rigid or semirigid pericardium is another characteristic of great importance.

A cardiac cylce is initiated by the filling of the sinus venosus. This filling is greatly accelerated during ventricular contraction when the pressure gradient between the larger veins steepens due to a *vis-à-fronte* effect set up by the downstroke in intrapericardial pressure.

Filling of the sinus venosus is a prolonged event occupying about half of the cardiac cycle length in teleosts (*Anguilla japonica*) (Chow and Chan 1975). The contraction of the sinus venosus raises its pressure above that of the posterior cardinal vein for about 30 per cent of the cardiac cycle, suggesting that, despite the lack of inflow valves at the junction of the ducts of Cuvier and the sinus venosus, some valve action must occur involving perhaps the flaccid walls of the venous inflow system to the heart.

Atrial diastole and filling go on longer than any other event of the cardiac cycle ($\simeq$ 52 per cent of the cardiac cycle in *Anguilla japonica*) for most of the time the atrium is being filled from the cardinal veins. Atrial filling terminates with sinus contraction which raises for a brief period (20-25 per cent of the cardiac cycle) its pressure in excess of both the post-cardinal venous and the atrial pressure. Most importantly, and different from birds and mammals, no filling of the ventricle occurs prior to atrial contraction. Throughout atrial diastole its pressure remains below the diastolic ventricular pressure (elasmobranchs, Sudak 1965a,b, and teleosts, Chow and Chan 1975) and the atrioventricular valves remain closed.

This has two important consequences. One, that ventricular filling, and thus its resulting end-diastolic volume, is not a prolonged passive process determined by a pressure head from the central veins. Rather the end-diastolic ventricular volume, and thus the subsequent ventricular stroke volume, is determined by an active contractile process, namely that of atrial contraction. The so-called preload on ventricular performance is thus represented by the atrial contraction and resulting pressures and not by the central venous pressure, as often reported also for fish (Farrell 1984). This implies that an elevated central venous pressure on the fish heart has importance for the ventricular end-diastolic volume only in so far as it may augment atrial stroke volume, which in turn is the sole agent for ventricular filling. The overlap in timing and duration of atrial contraction with ventricular filling makes the latter a much more brief event relative to the cardiac cycle length than is typical of birds and mammals.

During a considerable period of the diastole the ventricle thus remains isovolumetric. The duration of systole for both the atrium and ventricle in fish is longer than for higher vertebrates. This is only in part due to differences in temperature of the myocardium. A difference in the excitation-contraction process appears to exist. Thus atrial systole in *Anguilla japonica* lasted 31 per cent of a cardiac cycle length of 0.90 s, whereas ventricular systole lasted half the cardiac cycle. Duration of diastole will naturally vary with heart rate. At heart rates between 60-70 beats min^{-1}, ventricular diastole occupies 71-76 per cent of the cardiac cycle length in man, compared with 42 per cent in *Anguilla japonica* (Chow and Chan 1975). For other ectotherms such as the bullfrog, 43 per cent of the cycle length is ventricular diastole (Langille and Jones 1977), compared with 46 per cent for the Savannah monitor lizard (Burggren and Johansen 1982). The slow rate of the excitation-contraction process in ectotherms may have afforded a selection pressure for the reduced relative duration of diastole. In turn ventricular filling is limited to a shorter relative period and depends on the active contraction of the atrium. In this context it is of interest that atrial volume capacity in fish is similar to that of the ventricle, whereas in birds and mammals, which share with fish an in-series coupling of the systemic and gas-exchange vascular beds, the atrial volume capacity is much less than that of the ventricles. In accordance with this and the absence of valves at the atrial inflow side, only 30 per cent or less of the ventricular end-diastolic volume results from atrial contraction in homeotherms.

For comparative purposes myocardial (ventricular) contractility may be expressed by the rate of change of intraventricular pressure or perhaps also by the duration of isovolumetric contraction when no corrections are made for differences in myocardial temperature and afterload levels. In the former case ventricular contractility in *Anguilla japonica* (Chow and Chan 1975) is only about 13 per cent of that in man. The duration of ventricular isovolumetric contraction at nearly similar heart rates is 0.1 s for both the eel and man. For both species the contractility increases markedly with administration of noradrenaline or increased sympathetic activity. Whereas this for man, as expected, decreases the duration of systole, the only available data for a teleost heart (Chow and Chan 1975) indicate an increased ventricular systolic duration at an increased rate of change of intraventricular pressure following noradrenaline administration.

Increased cardiac contractility in the higher vertebrates exerts an important role as modulator of stroke volume by reducing the residual end-systolic ventricular volume. The very limited data available for the fish heart based on radio-cardiography suggest that in the resting fish the ventricle empties nearly completely, leaving no residual volume (Andresen, Ishimatsu, Johansen and Glass, 1986).

The function of the bulbus arteriosus is different in teleosts and elasmobranchs, which also is to be expected on account of the great difference in anatomy. Only the teleost bulbus will be discussed here since conditions in elasmobranchs are commendably well described by Satchell (1971). The bulbus arteriosus is a most important modulator of the haemodynamic events taking place in the short segment of ventral aorta between the heart and the gills. It has repeatedly been documented that the bulbus arteriosus exerts a profound pressure chamber or Windkessel effect in teleosts (Johansen, 1962; Stevens, Bennion, Randall and Shelton 1972). This implies that the ventricular ejectate and its associated kinetic energy are temporarily respectively stored and transferred to potential energy in the bulbus. The rebound of the highly elastic bulbus will greatly prolong the outflow from the heart into the ventral aorta. At normal heart rate, blood flow will be continuous in that vessel, a fact likely to have important implications for gas exchange in the gills. Jones, Langille, Randall and Shelton (1974) addressed the importance of ventral and dorsal aortic compliance in the light of the highly elastic bulbus arteriosus in teleost fish and related these qualities to the efficiency of branchial gas exchange. The bulbus arteriosus has been shown to be 32 times more distensible than the human aorta (Licht and Harris 1973). Based on *in vitro* pressure volume measurements of the bulbus, Priede (1976) calculated that elastic rebound of the bulbus may account for 25 per cent of neutral aortic blood flow. He also proposed that both structurally and functionally the teleost bulbus is distinctly different from the ventral aorta and is probably derived from cardiac rather than ventral aortic tissue. Based on radiographic analysis, Andresen *et al.* (1986) visualised the volume changes of the bulbus during the cardiac cycle. They showed that upon ventricular contraction the bulbus expands to accommodate nearly the entire ventricular ejectate. Rebound from the bulbus was so slow that its minimum size correlated with maximal filling of the ventricle, suggesting that ventral aortic flow is continuous.

Myocardial Morphology

Two types of vertebrate myocardial tissue occur. One consists of small bundles or trabeculae of cells which make up a spongy myocardium. The other is compact cardiac tissue, made up of densely packed cells. The spongy tissue has no or a sparse supply of coronary vessels. Thus as to gas exchange it has to rely mainly on venous blood passing through the tissue. Due to the design of circulation in fishes, this blood is variable in oxygen content and carbon dioxide pressure. Farrell (1984) has attempted to quantify the adequacy of the venous gas composition for the oxygen requirement of the spongy myocardium. The compact myocardium has richly developed coronaries carrying blood directly from the efferent gill circulation (Tota 1983).

The ratio of spongy to compact tissue varies from 1 to about 0.05 — the latter in mammals and birds. In ectothermic species it probably never attains values below 0.3 (Santer and Greer Walker 1980; Tota 1983). This ratio may also vary within a single species with age, etc. (Poupa, Gesser, Jonsson and Sullivan 1974).

The functional advantage of the simultaneous existence of the two tissues is not obvious. In the context of cellular energy expenditure, a large ventricular lumen with restricted trabeculation would seem favourable by offering a low frictional resistance to blood flow. At the same time, however, the cells have to develop a high wall tension in accordance with Laplace's law. This may be reduced in a spongy myocardium since this tissue may function as an assembly of small chambers expelling blood in parallel (Johansen 1965b).

Our knowledge about myocardial function is based almost exclusively on studies of a few mammalian and amphibian species, the latter frequently taken to represent all ectotherms. Specific features of the ectotherm cardiac cell including that of fishes may be missed, since generalisations are often based on features of the mammalian myocardium.

Electromechanical Coupling

The mechanical activity of the heart is governed by the cytoplasmic concentration of free Ca^{2+}, since myosin and actin only interact when Ca^{2+} is bound to the troponin molecules situated on the actin string. With an elevation of 'free' Ca^{2+} from below $10^{-7}M$ to about $10^{-6}M$, contractility turns from no to full activation (Carafoli 1985). The mechanisms behind this elevation, and the subsequent lowering of Ca^{2+} are still controversial. As for Na^+, the extra- and intracellular 'free' Ca^{2+} are far from equilibrium. Since the former is about $10^{-3}M$, the latter should, according to the Nernst equation, be about $0.3M$, but in reality it is about a million times lower (Chapman 1979). Therefore, Ca^{2+} will enter the cell passively, whereas it has to be removed by energy-dependent processes.

The resting cell membrane acts as an efficient barrier to Ca^{2+}. For the cardiac cell, however, this situation changes drastically during the action potential. A rapidly elicited and shortly lasting inward Na^+ flux causes the upstroke of the action potential in most excitable cells. In the cardiac cell, however, with its comparatively much prolonged action potential, most of the inward current is due to gates for Ca^{2+} and Na^+ opening immediately upon the upstroke, when the initial Na^+-current inactivates. The action potential terminates when these gates close and the K^+ efflux is enhanced.

The different processes involved have been reviewed recently by Noble (1984). Apart from the entrance just noted, associated with the slow inward current, Ca^{2+} may also enter by sarcolemmal exchange processes (Klitzner and Morad 1983; Noble 1984). In particular there is evidence that Ca^{2+} is coupled to Na^+ so that its intracellular activity adjusts passively to the Na^+ gradient and its extracellular activity. In addition, though, this exchange depends on the membrane potential, since more than 2 Na^+ seem to exchange for each Ca^{2+}. A depolarisation, then, would favour a net extrusion of positive charges, i.e. an influx of Ca^{2+} (Chapman 1979). It should, however, be noted that the stoichiometry and voltage dependence of this exchange is a matter of debate (Eisner and Lederer 1985).

Besides Ca^{2+} influx from the extracellular side, there may also be an intracellular redistribution of Ca^{2+} primarily from the sarcoplasmic reticulum (SR) to the cytoplasm. According to one hypothesis (Fabiato 1983) the release of Ca^{2+} from the SR is triggered in a dose-dependent way by the Ca^{2+} coming from the cell exterior during the action potential. If so, the SR should represent an amplifier. Studies on frogs (Fabiato 1983) and dogfish (Maylie, Nunzi and Morad 1980) suggest that SR contributes little to the cardiac electromechanical coupling in ectothermic vertebrates. In mammals and birds, however, the SR seems to deliver the main bulk of activator Ca^{2+}. The fact that the cardiac cells in general seem to be considerably smaller in ectothermic than in endothermic vertebrates (Kisch 1966; Helle 1983) is in keeping with this. A small cell is associated with a high surface-to-volume ratio and short distances of diffusion within the cell. Conceivably, then, it should depend less on structures like the SR for rapid changes of cytoplasmic Ca^{2+}. Meanwhile it should be recalled that heart rate and stroke volume vary greatly among species, such as in tuna and plaice. Together with this the SR influence on the electromechanical coupling may vary. This is suggested by biochemical as well as ultrastructural studies of hearts from different fishes (L. Dybro and H. Gesser unpublished; Santer 1974; McArdle and Johnston 1981; Helle 1983).

During repolarisation Ca^{2+} is removed from the cytoplasm causing an inactivation of contractility and a relaxation. In this process the Na^+-for-Ca^{2+} exchange is believed to lower intracellular 'free' Ca^{2+} towards a value in equilibrium with the Na^+ gradient, the extracellular Ca^{2+} and the membrane voltage, which now attains the negative resting value. There is also evidence for an outward Ca^{2+} transport dependent on a membrane-bound Ca^{2+} ATPase. In contrast to the Na^{+}-Ca^{2+} shift believed to carry the major part of Ca^{2+} leaving the cell, the Ca-ATPase seems to act as a high-affinity low-capacity system, finally reducing cytoplasm Ca^{2+} down to the resting level (Carafoli 1985). Similar to the situation during activation, the role of SR in accumulating Ca^{2+} during relaxation is unclear (Fabiato 1983).

Although the SR may be of minor importance in the undisturbed heart of fishes and other ectotherms, this may not be so during increased cardiac demand such as during physical activity, hypoxia or acidosis. The fish heart may be acidified due to increases in blood PCO_2 or lactic acid during exercise and stress, as well as during exposure to hypoxia in stagnant, trophic waters. Some hearts, like that of the flounder, seem to counteract the primarily negative inotropic effect of acidosis by a release of intracellular, probably mitchondrial, Ca^{2+} stores (Gesser and Poupa 1983). Mitochondria do not seem to participate directly in the beat-to-beat regulation of Ca^{2+}, whereas they may modulate the size of the Ca^{2+} pool involved in this regulation (Carafoli 1985).

An increase in resting tension of the myocardium should be expected upon a release of mitochondrial Ca^{2+} stores. Its absence suggests that the Ca^{2+} released is accumulated by the SR, in this way increasing the Ca^{2+} available for activation of contraction. Alternatively, the Ca^{2+} released may pass through the cytoplasm across the sarcolemma and thus offset the decreased sensitivity of myofilament to Ca^{2+}, which is believed to be a result of acidosis (Fabiato and Fabiato 1978). Assuming the original Ca^{2+} level to be below that necessary for activation, the effect of the increased cytoplasmic Ca^{2+} would only appear during activation.

A role of SR-like structures in the flounder heart has also been suggested in explaining the effects of osmotic changes (Gesser and Mangor-Jensen 1984).

Osmosis and Extracellular Ions

The osmotic pressure is one of several changes occurring in the extracellular space in fishes. Such a change may occur in the flounder by movement between waters of different salinity (Vislie and Fugelli 1975), as well as by exercise and stress (Holeton,

Neumann and Heisler 1983). Mainly, it is the NaCl concentration that changes. Therefore, the specific effects of Na^+, mainly on the Na^+-Ca^{2+} shift, should be expected to add to those of an osmotic change *per se*.

Extracellular K^+ is another variable known to increase during exercise due to its release from skeletal muscle (Holeton *et al.* 1983) and heart muscle (Klitzner and Morad 1983). Such an increase impoverishes the force development of the heart, particularly in ectotherms (Carmeliet 1977). With respect to the heart of rainbow trout, this effect apparently depends on beat frequency.

Remarkably, the heart muscle of teleosts tends to display a negative correlation between frequency and peak force, i.e. a negative staircase (Driedzic and Gesser 1985). For other vertebrates including elasmobranchs, but with the possible exception of cyclostomes, a positive staircase is the rule, at least in a lower frequency range. It is not clear why the teleost heart seems to be exceptional in this respect. However, at an elevated $[K^+]_o$ the staircase of the trout heart may become less negative or even reverse to positive (L. Hove-Madsen and H. Gesser unpublished). One hypothesis ascribes a positive staircase to an increase in intracellular Na^+ with frequency, whereby the Ca^{2+} influx during activation should be enhanced (Langer 1968). Possibly, the Na^+ concentration is high in the teleost cell, leaving little room for an effect of frequency. Conceivably, such a situation might be changed by a stimulation due to elevated $[K^+]_o$ of the Na^+, K^+ pump with a fall in $[Na^+]_i$ as a result.

Ruben and Bennett (1981) give evidence that extracellular Ca^{2+} may vary, particularly in ectotherms with a bony skeleton. The changes described would markedly affect heart performance. These results must remain inconclusive since they were not confirmed in another study (Andreasen 1985).

Metabolic Aspects

The few studies on contractile proteins of the fish heart suggest that these largely resemble those of the much more studied mammalian heart (Johnston 1983). However, many aspects need to be clarified. One topic concerns 'isoenzymes' of myosin and actin, and SR development in relation to heart performance. Fish would seem excellent subjects to examine for solving such problems since their cardiac performance is likely to vary greatly among species.

In accordance with the high demands on cardiovascular functions in mammals and birds, the myocardium relies almost solely on aerobic metabolism. The high and stable oxygen supply to the homeotherm heart is supported by an efficient gas exchange and a well-developed coronary system and high coronary blood flow. The question arises as to whether such a tissue oxygen supply could evolve in a water-breather. Water relative to air has a lower and more variable oxygen content, a low diffusing capacity for gases and a high viscosity, all making it less suitable as a breathing medium (Dejours 1975). The heart of fishes, in comparison with mammals and birds, has a lower work output and should be less limited in regard to energy-liberating pathways, and thus should be able to maintain its function during episodes of oxygen deficiency. Many fishes, such as eel and carp, frequently experience a hypoxic environment. It is also likely that some of the energy demand of the spongy heart tissue has to be covered anaerobically during exercise or other circumstances, when the quality of the venous blood flowing through the heart is oxygen deficient. In keeping with this, the anaerobic performance of heart tissue from different fish species falls within a range including values typical for mammals, but with the bulk of values far above. It relates fairly well to what would be expected from differences in living conditions (Gesser and Poupa 1974; Sidell 1983). This comparison may be oversimplified, however, in the light of observations on the rainbow trout heart. At a first sight, this heart appears to perform anaerobically as poorly as the rat heart. However, unlike the rat heart, its contractility is enhanced as much as during aerobic conditions, when $[Ca^{2+}]_o$ is elevated or adrenalin administered (Gesser, Andresen, Brams and Sund-Laursen 1982; Nielsen and Gesser 1983).

Biochemical studies also suggest a wide variation in cardiac metabolism among vertebrate groups. Fatty acids seem to be relatively more important in mammals and birds than in fishes, whereas the reverse is true for carbohydrates. In addition, the teleost heart appears unable to utilise ketone bodies (Sidell 1983), a substrate important in metabolism of the elasmobranch heart. On the other hand the elasmobranch heart seems to lack the capacity to utilise fatty acids (Sidell 1983). Further evidence for this is that the isolated skate heart was functioning well when supplied with glucose or ketone bodies. Fatty acids, however, had a negative inotropic effect (Driedzic and Hart 1984). It is noteworthy that elasmobranch blood, by lacking fatty acid binding proteins, seems to have a reduced capacity to carry fatty acids (Fellows and Hird 1981).

The heart of hagfish apparently prefers carbohydrates over fatty acids. Like the teleost heart it seems unable to utilise ketone bodies (Sidell 1983).

Cell size is probably of relevance in this context. Efficiency of the contractile machinery and rate of excitation seem to select for large cells. However, large cells with their low surface-to-volume ratio conceivably have a low capacity for exchanging material with the outside. This could be a problem in a situation like hypoxia, since the anaerobic, relative to the aerobic, metabolism is associated with a larger number of molecules transferred across the sarcolemma per ATP formed. In particular an efficient removal of lactic acid is important, since an accumulation of this anaerobic end product results in an acidification of the cell interior, and thus in a depressed mechanical performance.

With existing options, the dependence on exchange processes seems to be minimised with fatty acid catabolism. In this case the ATP molecules synthesised per carbon atom are maximal. Thus, such a catabolism should be suitable for large cardiac cells for which energy liberation is covered aerobically.

References

Andreasen, P. (1985) 'Free and Total Concentrations in Blood of Rainbow Trout, *Salmo gairdneri* during "Stress" Conditions', *Journal of Experimental Biology*, in press

Andresen, J.H., Ishimatsu, A., Johansen, K. and Glass, M.L. (1986) 'An Angiocardiographic Analysis of the Central Circulation in the Teleost, *Channa argus*' submitted for publication

Burggren, W.W. and Johansen, K. (1982) 'Ventricular Hemodynamics in the Monitor Lizard *Varanus exanthematicus*: Pulmonary and Systemic Pressure Separation', *Journal of Experimental Biology*, 96, 343-54

Carafoli, E. (1985) 'The Homeostasis of Calcium in Heart Cells', *Journal of Molecular and Cellular Cardiology*, 17, 203-12

Carmeliet, E. (1977) 'Repolarization and Frequency in Cardiac Cells', *Journal de Physiologie (Paris)*, 73, 903-23

Chapman, R.A. (1979) 'Excitation Contraction Coupling in Cardiac Muscle', *Proceedings in Biophysics and Molecular Biology*, 35, 1-52

Chow, P.H. and Chan, D.K.O. (1975) 'The Cardiac Cycle and the Effects of Neurohumors on Myocardial Contractility in the Asiatic Eel, *Anguilla japonica*', *Comparative Biochemistry and Physiology*, 52C, 41-5

Dejours, P. (1975) *Principles of Comparative Respiratory Physiology*, North Holland/American Elsevier

Driedzic, W.R. and Gesser, H. (1985) 'Ca^{2+} Protection from the Negative Inotropic Effect of Contraction Frequency on Teleost Hearts', *Journal of Comparative Physiology, 156*, 135-42

Driedzic, W.R. and Hart, T. (1984) 'Relationship between Exogenous Fuel Availability and Performance by Teleost and Elasmobranch Hearts', *Journal of Comparative Physiology, 154*, 593-9

Eisner, D.A. and Lederer, W.J. (1985) 'Na-Ca Exchange: Stoichiometry and Electrogenicity', *American Journal of Physiology, 248*, C189-202

Fabiato, A. (1983) 'Calcium-induced Release of Calcium from the Cardiac Sarcoplasmic Reticulum', *American Journal of Physiology, 245*, C1-14

Fabiato, A. and Fabiato, F. (1978) 'Effects of pH on the Myofilaments and the Sarcoplasmic Reticulum of Skinned Cells from Cardiac and Skeletal Muscles', *Journal of Physiology, 276*, 233-55

Farrell, A.P. (1984) 'A Review of Cardiac Performance in the Teleost Heart: Intrinsic and Humoral Regulation', *Canadian Journal of Zoology, 62*, 523-36

Fellows, F.C.I. and Hird, F.J.R. (1981) 'Fatty Acid Binding Proteins in the Serum of Various Animals', *Comparative Biochemistry and Physiology, 68B*, 83-7

Gesser, H. and Mangor-Jensen, A. (1984) 'Contractility and ^{45}Ca Fluxes in Heart Muscle of Flounder at a Lowered Extracellular NaCl Concentration', *Journal of Experimental Biology, 109*, 201-7

Gesser, H. and Poupa, O. (1974) 'Relations between Heart Muscle Enzyme Pattern and Directly Measured Tolerance to Anoxia', *Comparative Biochemistry and Physiology, 48*, 97-104

Gesser, H. and Poupa, O. (1983) 'Acidosis and Cardiac Muscle Contractility: Comparative Aspects', *Comparative Biochemistry and Physiology, 76A*, 559-67

Gesser, H., Andresen, P., Brams, P. and Sund-Laursen, J. (1982) 'Inotropic Effects of Adrenaline on the Anoxic or Hypercapnic Myocardium of Rainbow Trout and Eel', *Journal of Comparative Physiology, 147*, 123-8

Helle, K. (1983) 'Structures of Functional Interest in the Myocardium of Lower Vertebrates', *Comparative Biochemistry and Physiology, 76A*, 447-52

Holeton, G.F., Neumann, P. and Heisler, N. (1983) 'Branchial Ion Exchange and Acid-Base Regulation after Strenuous Exercise in Rainbow Trout (*Salmo gairdneri*)', *Respiration Physiology, 51*, 303-18

Johansen, K. (1962) 'Cardiac Output and Pulsatile Aortic Flow in the Teleost, *Gadus morhua*', *Comparative Biochemistry and Physiology, 7*, 169-74

Johansen, K. (1965a) 'Dynamics of Venous Return in Elasmobranch Fishes', *Hvalradets Skrifter, 48*, 94-100

Johansen, K. (1965b) 'Cadiovascular Dynamics in Fishes, Amphibians and Reptiles', *Annals of the NY Academy of Science, 127*, 414-42

Johansen, K. and Hanson, D. (1967) 'Hepatic Vein Sphincters in Elasmobranchs and their Significance in Controlling Hepatic Blood Flow', *Journal of Experimental Biology, 46*, 195-203

Johnston, I.A. (1983) 'Comparative Studies of Contractile Proteins from the Skeletal and Cardiac Muscles of Lower Vertebrates', *Comparative Biochemistry and Physiology, 76A*, 439-46

Jones, D.R., Langille, B.W., Randall, D.J. and Shelton, G. (1974) 'Blood Flow in the Dorsal and Ventral Aorta of the Cod, *Gadus morhua*', *American Journal of Physiology, 226*, 90-5

Kisch, B. (1966) 'The Ultrastructure of the Myocardium of Fishes', *Experimental Medical Surgery, 24*, 220-7

Klitzner, T. and Morad, M. (1983) 'Excitation-Contraction Coupling in Frog Ventricle. Possible Ca^{2+} Transport Mechanisms', *Pflügers Archiv, 398*, 274-83

Langer, G. (1968) 'Ion Fluxes in Cardiac Excitation and Contraction and their Relation to Myocardial Contractility', *Physiological Reviews, 48*, 708-57

Langille, B.L. and Jones, D.R. (1977) 'Dynamics of Blood Flow through the Hearts and Arterial Systems of Anuran Amphibia', *Journal of Experimental Biology, 68*, 1-17

Licht, J.H. and Harris, N.S. (1973) 'The Structure, Composition and Elastic Properties of the Teleost Bulbus Arteriosus in the Carp', *Comparative Biochemistry and Physiology, 46A*, 699-708

McArdle, H.J. and Johnston, I.A. (1981) 'Ca^{2+}-uptake by Tissue Sections and Biochemical Characteristics of Sarcoplasmic Reticulum Isolated from Fish Fast and Slow Muscles', *European Journal of Cell Biology, 25*, 103-7

Maylie, J., Nunzi, M.G. and Morad, M. (1980) 'Excitation-Contraction Coupling in Ventricular Muscle of Dogfish (*Squalus acanthias*)', *The Bulletin of the Mount Desert Island Biological Laboratory, Salisbury Cove, Maine, 19*, 84-7

Nielsen, K.E. and Gesser, H. (1983) 'Effects of [Ca^{2+}]$_o$ on Contractility in the Anoxic Cardiac Muscle of Mammal and Fish', *Life Sciences, 32*, 1437-42

Noble, D. (1984) 'The Surprising Heart: a Review of Recent Progress in Cardiac Electrophysiology', *Journal of Physiology, 353*, 1-50

Poupa, O., Gesser, H., Jonsson, S. and Sullivan, L. (1974) 'Coronary Supplied Compact Shell of Ventricular Myocardium in Salmonids: Growth and Enzyme Pattern', *Comparative Biochemistry and Physiology, 48A*, 85-95

Priede, I.G. (1976) 'Functional Morphology of the Bulbus Arteriosus of Rainbow Trout (*Salmo gairdneri*)', *Journal of Fish Biology, 9*, 209-16

Ruben, J.A. and Bennett, A.F. (1981) 'Intense Exercise, Bone Structures and Blood Calcium Levels in Vertebrates', *Nature, 291*, 411-13

Santer, R.M. (1974) 'The Organization of the Sarcoplasmic Reticulum in Teleost Ventricular Myocardial Cells', *Cell and Tissue Research, 151*, 355-402

Santer, R.M. and Greer Walker, M. (1980) 'Morphological Studies on the Ventricle of Teleost and Elasmobranch Hearts', *Journal of Zoology (London), 190*, 259-72

Satchell, G.H. (1971) *Circulation in Fishes*, Cambridge University Press, London

Shabetai, R., Abel, D.C., Graham, J.B., Bhargava, V., Keyes, R.S. and Witztum, K. (1985) 'Function of the Pericardium and Pericardio-peritoneal Canal in Elasmobranch Fishes', *American Journal of Physiology, 248 (Heart Circulation Physiology, 17)*, H198-207

Sidell, B.D. (1983) 'Cardiac Metabolism in the Myxinidae: Physiological and Phylogenetic Considerations', *Comparative Biochemistry and Physiology, 76A*, 495-506

Stevens, E.D., Bennion, G.R., Randall, D.J. and Shelton, G. (1972) 'Factors Affecting Arterial Blood Pressures and Blood Flow from the Heart in Intact, Unrestrained Ling Cod, *Ophiodon elongatus*', *Comparative Biochemistry and Physiology, 43*, 681-95

Sudak, F.N. (1965a) 'Intrapericardial and Intracardiac Pressures and the Events of the Cardiac Cycle in *Mustelus canis* (Mitchell)', *Comparative Biochemistry and Physiology, 14*, 689-705

Sudak, F.N. (1965b) 'Some Factors Contributing to the Development of Subatmospheric Pressure in the Heart Chambers and Pericardial Cavity of *Mustelus canis* (Mitchell)', *Comparative Biochemistry and Physiology, 15*, 199-215

Tota, B. (1983) 'Vascular and Metabolic Zonation in the Ventricular Myocardium of Mammals and Fishes', *Comparative Biochemistry and Physiology, 76A*, 423-37

Vislie, T. and Fugelli, K. (1975) 'Cell Volume Regulation in Flounder (*Platichthys flesus*) Heart Muscle Accompanying an Alteration in Plasma Osmolarity', *Comparative Biochemistry and Physiology, 52A*, 415-18

5   CONTROL OF GILL BLOOD FLOW

Stefan Nilsson

The arrangement of the branchial vasculature in fishes is very complex, and only during the last ten years have there been major advances in our understanding of the microanatomy of the gill vasculature. The main reason for the increase in knowledge is the introduction of vascular casting techniques, which allow a detailed study of the vascular spaces by binocular microscope or scanning electron microscope (Gannon, Campbell and Randall 1973; Laurent and Dunel 1976; Cooke 1980; Cooke and Campbell 1980; see also Laurent 1984).

The mechanisms of blood-flow control in fish gills can only be clarified if the anatomical basis is well established. The overall branchial vascular resistance of perfused gills can be altered by many substances, notably catecholamines (Krawkow 1913; Nilsson 1984a), but the knowledge of the structure and function of the branchial vasomotor innervation and the sites of action of nerves and humoral agents is still fragmentary.

The aim of this chapter is to summarise some studies undertaken to elucidate the mechanisms responsible for blood flow control in fish gills. Since the bulk of available information deals with teleosts, the discussion will be focused mainly on this group of fish. For in-depth accounts of general, vascular and nervous anatomy and physiology of fish gills, the reader is referred to Hughes (1984), Laurent (1984), Nilsson (1984a) and Randall and Daxboeck (1984).

Anatomical Considerations

The gill arches carry two rows of filaments arranged in the respiratory water current to maintain a continuous 'gill curtain'. The position of the filaments is controlled by striated adductor and abductor muscles, which can rapidly alter the angle between the two filament rows, and there is now also good evidence for the presence of smooth abductor muscles innervated by adrenergic nerve fibres from the autonomic nervous system (Laurent 1984; Nilsson 1985).

Blood ejected from the heart into the ventral aorta runs in the four pairs of afferent branchial arteries within the gill arches (Figures 5.1 and 5.2). Afferent filamental arteries, in turn, give off the afferent lamellar arterioles entering the gill lamellae. The gill lamellae consist of two tissue sheets held together by specialised endothelial cells, the pillar cells, and blood flows through the lamellae in the opposite direction to the respiratory water passing between the lamellae (countercurrent flow). After passing the lamellae, the blood is re-collected into the efferent branchial arteries, via the efferent lamellar arterioles and efferent filamental

Figure 5.1: Simplified Diagrammatic Representation of the Vascular Arrangement in the Gill Region of the Cod, *Gadus morhua*. Blood is ejected from the heart into the ventral aorta, which gives off the four pairs of afferent branchial arteries (ABA) entering the gill arches. Blood leaving the gills via the efferent branchial arteries (EBA) enters the suprabranchial artery (SUPA) which gives off the carotid artery (CA) anteriorly, and the dorsal aorta (DA) and coeliaco-mesenteric artery (CMA) posteriorly. Blood leaving the gills in the branchial veins (BV) may enter the anterior cardinal vein (ACV) dorsally, or the inferior jugular vein (IJV) ventrally. The venous blood is collected in the duct of Cuvier (DC) and enters the sinus venosus of the heart. Arrows indicate the direction of blood flow

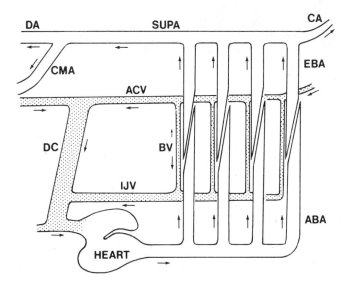

Figure 5.2: A Simplified Summary of the Main Features of the Vascular Anatomy of the Gills of the Atlantic Cod, *Gadus morhua*. The filamental cartilage rods and the 'nutritive' vasculature (except the central venous system and the subepithelial veins) are not shown in the figure. The size of the blood vessels is slightly exaggerated for clarity. *Abbreviations used*: ABA, afferent branchial artery; AFA, afferent filamental artery; ALa, afferent lamellar arteriole; BN, branchial nerve; BV, branchial veins; CVS, central venous system; EBA, efferent branchial artery, EFA, efferent filamental artery; ELa, efferent lamellar arteriole; Fil, gill filaments; GA, gill arch bone; Lam, lamella; SuV, subepithelial vein. Arrows indicate the direction of blood flow

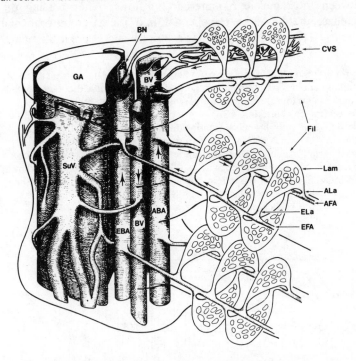

arteries (Figure 5.2). The efferent branchial arteries enter dorsally the suprabranchial artery on each side of the skull. The two suprabranchial arteries join to form the dorsal aorta (Figure 5.1).

The blood pathway through the gills described so far (afferent arteries — lamellae — efferent arteries) is known as the *arterio-arterial pathway*, and is of particular interest for the respiratory function of the gill. There is also an *arterio-venous pathway*, which is of particular interest for the osmoregulatory functions of gills, since the epithelium of the gill filament, which contains most of the ion-transporting chloride cells, is in contact with the blood entering the central venous system at the core of the filament (see Laurent 1984). The arterio-venous pathway consists of two major parts: arterio-venous anastomoses, which let blood from the filamental arteries directly enter the central filamental venous system (CVS, see Figure 5.2), and a system of 'nutritive' vascular beds of the gill tissues, not essentially different from other systemic vascular beds.

Arterio-venous anastomoses occur, at least in most teleosts studied, between the efferent filamental artery and central venous system. In some species, e.g. the eel, there are additional anastomoses between the afferent filamental artery and the central venous system (Laurent and Dunel 1976; Vogel 1978; Laurent 1984).

The 'nutritive' vasculature of the gill arch and filament is derived from the efferent filamental and branchial arteries, and forms 'normal' systemic vascular beds in the tissues of the gills (Cooke and Campbell 1980; Laurent 1984). In addition to such 'normal' systemic vascular beds, the presence of a secondary arterial system in fish gills has been proposed (Vogel 1981).

The venous blood from the vascular beds of the gill arch is collected either directly into the branchial veins, or into large subepithelial veins. Whether the various venous compartments of the gills should be regarded as 'venous', 'veno-lymphatic' or 'lymphatic' has been the cause of controversy (see Cooke and Campbell 1980; Vogel 1981; Laurent 1984). For simplicity, the term 'venous' is used throughout this chapter. The central venous system of the filaments is in some species (e.g. eel) arranged as a single sinus whereas in others (e.g. cod; Figure 5.2) there is instead a plexus of interconnected vessels. The venous blood from filamental structures ends up in the central venous system, which is also connected to the branchial veins.

The branchial veins drain dorsally into the anterior cardinal vein and ventrally into the inferior jugular vein. The dorsal connections in the cod, *Gadus morhua* (Figures 5.1, 5.2) are very narrow, but in other species (e.g. eel; Rowing 1981) there may be a substantial venous drainage into the anterior cardinal veins.

Vascular Innervation

The gills receive a very rich innervation by the branchial branches of the glossopharyngeal (IX) and vagus (X) nerves. The bulk of the fibres running in these nerves are believed to be sensory (proprioceptor, nociceptor, baroreceptor and chemoreceptor fibres), but there are also autonomic fibres and motor fibres to the branchial muscles (e.g. filamental abductor and adductor muscles; see Nilsson 1984a).

Autonomic vasomotor fibres to the gill vasculature can be of two types; cranial autonomic ('parasympathetic') fibres, which run in the cranial nerves, and spinal autonomic ('sympathetic') fibres. The latter originate from the sympathetic chains and join the branchial nerves via grey *rami communicantes* near the exit of the cranial nerves from the skull (Nilsson 1983, 1984a). So far, only the presence of the 'classical' autonomic nerve types (cholinergic and adrenergic nerves) has been investigated, but it is clear that a large number of putative non-adrenergic, non-cholinergic (NANC) neurotransmitters occur in the autonomic nervous system (see Nilsson 1983, 1984a) and may be present in the gills.

Nerve terminals, indicative of a functional innervation, have been described by histochemical and electron microscopical techniques at several potential effector sites in the gill vasculature. c-Type nerve terminals, suggestive of a cholinergic innervation, have been described in a vascular sphincter in the efferent filamental arteries, near the base of the filament (Dunel and Laurent 1980). The bulk of evidence at the moment seems to favour the view that these fibres are cholinergic vasoconstrictor fibres of vagal origin (see later).

In addition to these sphincters, an innervation of the efferent arterio-venous anastomoses of *Tilapia* has been described (Vogel, Vogel and Schlote 1974). A claim that the pillar cells of the lamellae are innervated (Gilloteaux 1969) lacks confirmation (Laurent and Dunel 1980).

Adrenergic nerve fibres, observed by fluorescence histochemistry, have been described in the gills of trout (*Salmo gairdneri, S. trutta*; Donald 1984), cod (*Gadus morhua*; Figure 5.3A-D) and the air-breathing teleost *Channa argus* (Ishimatsu, Johansen and Nilsson 1986). In both *Salmo* species, there is a rich innervation of nutritive vessels in the filament, and there are also adrenergic fibres in the afferent and efferent lamellar arterioles. In *Salmo trutta* there are, in addition, adrenergic fibres in the walls of the afferent and efferent branchial arteries, and at the bases of the efferent filamental arteries (Donald 1984). In the cod, there are

Figure 5.3: Fluorescence Histochemistry (Falck-Hillarp Technique) Showing Adrenergic Nerve Fibres in the Gills of the Atlantic Cod, *Gadus morhua*. *A:* Single weakly fluorescent varicose nerve fibres (arrows) running at the base of afferent lamellar arterioles (a) along the afferent filamental artery (not shown). The fluorescence of the afferent lamellar arterioles is due to a faint autofluorescence of the vessel wall. *B:* Transverse section through the branchial nerve within the gill arch. Note large autofluorescent (non-adrenergic) ganglion cells (gc) and adrenergic fibres within the branchial nerve (arrows). *C:* A sparse plexus of varicose adrenergic nerve fibres in the adductor muscle of the cod gill. *D:* Dense adrenergic nerve plexus in the longitudinally cut smooth abductor muscle from cod gills. All preparations are taken from the 3rd pair of gill arches. Calibration bars in all figures = 50 μm

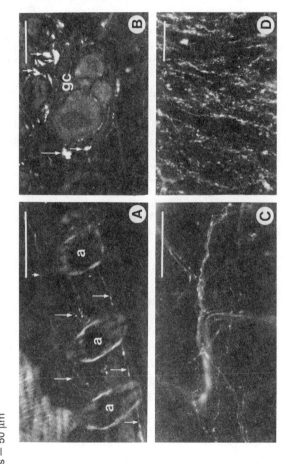

scarce varicose adrenergic fibres at the bases of the afferent, but not the efferent, lamellar arterioles (Figure 5.3A), and along the nutritive vessels of the filament (Figure 5.3C). Adrenergic fibres are present in the branchial nerve within the gill arch, but the ganglion cells observed appear to be solely non-adrenergic (Figure 5.3B). The structure in the cod gill with the densest adrenergic innervation is the smooth abductor muscle joining the ends of the filamental cartilage rods within the gill arch (Figure 5.3D; see also Laurent 1984; Nilsson 1985). In *Channa* an adrenergic innervation of nutritive blood vessels in the core of the filament was observed (Ishimatsu *et al.* 1986). This situation is similar to that in other teleost species.

It would seem that the evidence obtained so far from histochemical and electron microscopical work indicates an innervation, at least in some species, of the lamellar arterioles, the base of the efferent filamental artery and, especially, the nutritive vasculature of the gill.

Nervous Control of the Gill Vasculature

Attempts to provoke changes in the vascular resistance of fish gills by electrical stimulation of the branchial nerve have been made by Nilsson (1973) and later by Pettersson and Nilsson (1979). These experiments on the cod demonstrated the presence of both cholinergic and adrenergic vasomotor fibres to the branchial vasculature, but the sites of action could not be determined. In later experiments, both the venous (inferior jugular vein) and the arterial (dorsal aortic) outflow and the inflow counterpressure (ventral aortic pressure) was monitored in cod gills perfused at constant flow. In these experiments the gill apparatus of one side only was perfused, and the sympathetic chain on the same side could be stimulated electrically. Thus the responses of the gill vasculature in the experiments only reflected the vasomotor innervation by autonomic nerve fibres of spinal autonomic ('sympathetic') origin (Nilsson and Pettersson 1981).

The results from the experiments show that, upon electrical stimulation of the sympathetic chains, there is an increase in the arterio-arterial flow and a decrease in the arterio-venous flow (Figure 5.4). The response could be reversed (counterpressure and venous outflow) or strongly reduced (arterial outflow) by the α-

Figure 5.4: Recordings of Afferent Counterpressure (P_i), Arterial ($\dot{Q}_a$) and Venous ($\dot{Q}_v$) Flow in a Right-side Gill Apparatus of the Atlantic Cod, *Gadus morhua*, Perfused at Constant Flow. Inflow pressure was measured by a pressure transducer and outflows by photoelectric drop-counters. At points indicated the sympathetic nerve supply to the gills was stimulated electrically (1 ms pulse duration at 10 Hz and 8 V for 1 min with 8 min intervals). Upper tracings: effect of phentolamine 10^{-6}M on the vascular responses to nerve stimulation. Note the slight reversal in P_i and $\dot{Q}_v$ after addition of the drug. Reproduced with permission from Nilsson and Pettersson (1981). Copyright Springer-Verlag 1981. Lower tracings: effect of propanolol 3×10^{-6}M on the vascular responses to nerve stimulation. Note a slight increase in basal P_i and reduction in the nervously induced arterial flow increase. P_i in both sets of tracings is expressed in kilopascals (kPa) and flows in drops per minute

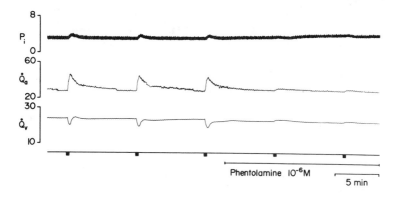

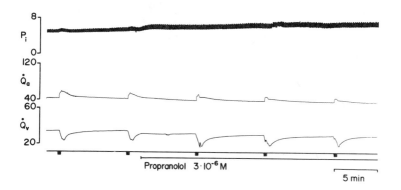

adrenoceptor antagonist phentolamine, but was only partially reduced by the β-adrenoceptor antagonist propranolol. The result can be explained simply by assuming that the 'sympathetic' innervation controls the nutritive vasculature by adrenergic vasoconstrictor fibres, a suggestion that is corroborated by the histochemical evidence. Constriction of the nutritive vascular beds would reduce the venous outflow from the gills, and simultaneously redirect the flow to the arterio-arterial pathway. A closing of systemic vascular leaks by the nerve stimulation, an experimental artefact, will also affect the outcome.

The presence of vascular elements where the 'sympathetic' innervation produces vasodilation is indicated by the reversal of the inflow counterpressure response and the venous outflow response after phentolamine. The increase in inflow counterpressure after propranolol suggests the involvement of β-adrenoceptors in the vasodilatory responses (Figure 5.4).

Retrograde Perfusion

In these experiments an isolated gill arch is perfused via the efferent branchial artery (retrograde perfusion), and the perfusion back-pressure is monitored. The outflow from the gill is almost entirely venous at the perfusion pressure used ($c.$ 3.5 kPa, similar to dorsal aortic blood pressure in the cod), and no more than a trickle of perfusion fluid will pass through the lamellae into the afferent arterial system. Changes in the perfusion back-pressure under these experimental conditions will thus reflect changes in the vascular resistance in the nutritive vasculature only.

Stimulation of the branchial nerve in such a preparation demonstrated the vasoconstrictor innervation of the gill vasculature previously described (Figure 5.5; Pettersson and Nilsson 1979). Since almost no flow through the arterio-arterial pathway existed in the preparation, the result must be interpreted in terms of a vasoconstrictor innervation of the arterio-venous pathway.

Cutting the filaments during perfusion only slightly reduced the back-pressure during maintained constant flow, and the constrictor response to branchial nerve stimulation remained virtually unchanged after the cut (Figure 5.5). There are two ways to interpret this result. The first is to assume that the major resistance along the efferent arterial pathway is at the 'Dunel–Laurent

Figure 5.5: Effects of Branchial Nerve Stimulation on Perfusion Back-pressure in an Isolated Cod Gill Arch Perfused at Constant Flow through the Efferent Branchial Artery (retrograde perfusion). As indicated by the arrow, both rows of filaments on the gill arch were cut to about half their original length. Note the slight reduction in perfusion back-pressure, and the persistence of the nerve response. Stimulation of the branchial nerve as indicated was carried out with 50s trains of pulses at 10Hz, 1 ms pulse duration and 8V

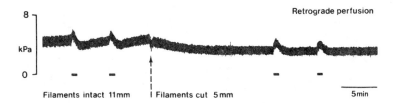

sphincters' at the bases of the efferent filamental arteries, and that the branchial nerve exerts its vasomotor control by constriction of these sphincters.

The other possibility is that the cutting of the filaments, notably the cutting of the efferent filamental arteries, will not produce a major leakage, but that the arteries constrict and the flow profile through the arterio-venous pathway remains virtually intact despite the cut. Observations during tip transducer measurements of intravascular blood pressure in the filamental arteries of the cod favour the latter explanation (Riegel and Nilsson, unpublished). In these preliminary experiments it was noted (a) that the dorsal aortic blood pressure and the efferent filamental arterial blood pressure, at least during the conditions of the experiments (i.e. anaesthetised fish), were equal, refuting the presence of a major resistance at the base of the efferent filamental arteries; and (b) that both the afferent and the efferent filamental arteries responded to piercing with the transducer micropipette tip by constriction along part of the artery.

It would thus seem, based on the experiments so far conducted, that the vasomotor innervation of cod gills is chiefly one of the nutritive vasculature, and that a nervous control of the arterio-arterial pathway, although evident in some experiments, is of less importance.

Humoral Control of the Gill Vasculature

In teleost fish the catecholamines (adrenaline and noradrenaline) are stored in and released from the chromaffin tissue lining the posterior cardinal veins inside the head kidney. Catecholamines released from this tissue into the bloodstream will rapidly reach the gills, and may serve as vasoregulatory agents (Wahlqvist 1981; Nilsson 1984b). In addition to the catecholamines, several other hormones could affect gill functions, including blood-flow patterns, but the physiological significance of observations of vasomotor activity of, for example, peptides remains to be elucidated (cf. Nilsson 1984a).

Numerous authors have described vasomotor effects of adrenaline and related substances in perfused fish gills (for references, see, for example, Nilsson, 1984a). The most commonly observed effect of adrenalin is a β-adrenoceptor-mediated vasodilation of the arterio-arterial pathway, often preceded by a transient α-adrenoceptor-mediated vasoconstriction. Under some circumstances, the constrictor response dominates (e.g. Wahlqvist 1981; Pärt, Kiessling and Ring 1982; Pettersson 1983).

Comparison of the concentration/response curve for adrenaline in the perfused gill arch on one hand, and the plasma concentration of adrenaline on the other, suggests that at least under some conditions, such as severe physical disturbance or hypoxia (= 'stress'), the humoral catecholamines affect the branchial vasculature of rainbow trout and cod (Nakano and Tomlinson 1967; Wood 1974; Wahlqvist 1980; Wahlqvist and Nilsson 1980). It should be noted that sustained swimming exercise, or even exhaustive exercise, does not produce an increase in the plasma levels of adrenaline in the rainbow trout or cod sufficient to more than marginally affect the gill vasculature (Primmett, Randall, Mazeaud and Boutilier 1986; Axelsson and Nilsson, to be published; see also discussion by Butler, Chapter 6 of this volume).

Adrenaline affects the flow in perfused cod gills to favour the arterio-arterial pathway, and since the β-adrenoceptor agonist isoprenaline lacks this effect, it was concluded that the redistribution of flow was due to to an α-adrenoceptor-mediated effect, possibly similar to that caused by adrenergic nerves (Nilsson and Pettersson 1981). However, an overall decrease in the branchial vascular resistance was observed during perfusion with isoprenaline, demonstrating the presence of β-adrenoceptors in the gill vasculature.

It can be concluded that catecholamines, released from chromaffin tissue, under some circumstances may reach plasma concentrations high enough to affect the branchial vasculature. The dominant effect of adrenaline is, with few exceptions, vasodilation of the arterio-arterial pathway, but an α-adrenoceptor-mediated constriction of the arterio-venous pathway is also present (Nilsson and Pettersson 1981).

Consequences of Changes in the Gill Blood Flow Patterns

It is now clear that the gills serve at least two major functions in fish: that of respiratory gas exchange, and that of osmo- and ion regulation. The respiratory advantage of a large functional surface area where gas exchange can take place is associated with an osmoregulatory disadvantage due to ionic loss/water influx (freshwater teleosts) or water loss (marine teleosts). Regulation of the functional surface area will balance this 'respiratory osmoregulatory compromise' at different oxygen demands, e.g. during exercise (Randall, Baumgarten and Malyusz 1972; Wood and Randall 1973).

There are basically two ways in which the functional surface area could be controlled:

(1) Some of the lamellae can be closed off by constriction of the afferent lamellar arterioles during rest, and opened during periods of increased oxygen demand (lamellar recruitment; Booth 1978, 1979; Holbert, Boland and Olson 1979);
(2) The blood can pass through the lamellae with little contact with the surrounding water, e.g. via channels at the base of the lamellae. These channels are embedded in the body of the filament, and thus lack the very intimate contact with the respiratory water (Farrell, Sobin, Randall and Crosby 1980; Pärt, Tuurala, Nikinmaa and Kiessling 1984).

The validity of these two models, alone or in combination, remains to be elucidated further, and much work is needed before the sites of action of nerves and humoral agents can be pinpointed exactly.

Lamellar recruitment can be elicited by adrenaline administration, and this effect can be observed in an anaesthetised fish under a powerful binocular microscope. In the cod, stagnant red blood cells are present in individual lamellae and after injection of adrenaline the lamellae are rapidly flushed and blood flows through all observed lamellae (S. Nilsson, unpublished). The observed effect can be due to a number of events, alone or in combination:

(1) Increased cardiac stroke volume and ventral aortic pressure causing a reduced branchial vascular resistance due to the compliance of the gill vasculature (Farrell, Daxboeck and Randall 1979).
(2) Dilation of the afferent lamellar arterioles (Farrell 1980).
(3) Constriction of efferent lamellar arterioles which would produce an increased intralamellar 'opening pressure'. This type of constriction may also occur as a direct effect on the blood vessels of hypoxia (Pettersson 1983; Pettersson and Johansen 1982).
(4) Constriction of the 'Dunel-Laurent sphincters' at the bases of the efferent filamental arteries. However, if this effect is of significance for the blood-flow distribution, it is likely to be due to a cholinergic innervation of these sphincters, rather than an adrenergic effect.
(5) A general increase in the efferent ('systemic') blood pressure which, as in (3) and (4) above, would increase intralamellar 'opening pressure'.

Changes in the intralamellar blood-flow pattern, i.e. a shunting of the blood through the basal channels of the lamellae (Farrell *et al.* 1980; Pärt *et al.* 1984) could be due simply to an effect of the intralamellar blood pressure. An increase in the intralamellar blood pressure, due to any combination of the events suggested above, would tend to increase the lamellar thickness and thus blood flow through the central part of the lamella. Reduced intralamellar pressure, on the other hand, may cause narrowing of the central lamellar blood space and thus favour blood flow in the basal channels (cf. Farrell *et al.* 1980).

It can thus be concluded that changes in the blood-flow profile in fish gills, resulting from the activity of neuronal and humoral control systems, can affect the functional surface area of the gills and thus the transfer of gases, water and solutes across the gill membranes.

Acknowledgements

Many thanks to Ms Aino Falck-Wahlström and Ms Birgitta Vallander for help with the illustrations. Our own research on fish gills is currently supported by grants from the Swedish Natural Science Research Council.

References

Booth, J.H. (1978) 'The Distribution of Blood in the Gills of Fish: Application of a New Technique to Rainbow Trout (*Salmo gairdneri*), *Journal of Experimental Biology*, *73*, 119-29

Booth, J.H. (1979) 'Circulation in Trout Gills: the Relationship between Branchial Perfusion and the Width of the Lamellar Blood Space', *Canadian Journal of Zoology*, *57*, 2183-5

Cooke, I.R.C. (1980) 'Functional Aspects of the Morphology and Vascular Anatomy of the Gills of the Endeavour Dogfish, *Centrophorus scalpratus* (McCullough) (Elasmobranchii: Squalidae)', *Zoomorphologie*, *94*, 167-83

Cooke, I.R.C. and Campbell, G. (1980) 'The Vascular Anatomy of the Gills of the Smooth Toadfish, *Torquiginer glaber* (Teleostei: Tetraodontidae)', *Zoomorphologie*, *94*, 151-66

Donald, J. (1984) 'Adrenergic Innervation of the Gills of Brown and Rainbow Trout, *Salmo trutta* and *S. gairdneri*', *Journal of Morphology*, *182*, 307-16

Dunel, S. and Laurent, P. (1980) 'Functional Organization of the Gill Vasculature in Different Classes of Fish', in B. Lahlou (ed.) *Jean Maetz Symposium: Epithelial Transport in the Lower Vertebrates*, Cambridge University Press, London, pp. 37-58

Farrell, A.P. (1980) 'Gill Morphometrics, Vessel Dimensions, and Vascular Resistance in Ling Cod, *Ophiodon elongatus*', *Canadian Journal of Zoology*, *58*, 807-18

Farrell, A.P., Daxboeck, C. and Randall, D.J. (1979) 'The Effect of Input Pressure and Flow on the Pattern and Resistance to Flow in the Isolated Perfused Gill of a Teleost Fish', *Journal of Comparative Physiology*, *133*, 233-40

Farrell, A.P., Sobin, S.S., Randall, D.J. and Crosby, S. (1980) 'Intralamellar Blood Flow Patterns in Fish Gills', *American Journal of Physiology*, *239*, R428-36

Gannon, B.J., Campbell, G. and Randall, D.J. (1973) 'Scanning Electron Microscopy of Vascular Casts for the Study of Vessel Connections in a Complex Vascular Bed, the Trout Gill', *Proceedings from the Annual Meeting, Electron Microscopical Society of America*, *31*, 442-3

Gilloteaux, J. (1969) 'Note sur l'innervation des branchies chez *Anguilla anguilla* L.', *Experientia*, *25*, 270

Holbert, P.W., Boland, E.J. and Olson, K.R. (1979) 'The Effect of Epinephrine and Acetylcholine on the Distribution of Red Cells within the Gills of the Channel Catfish (*Ictalurus punctatus*)', *Journal of Experimental Biology*, *79*, 135-46

Hughes, G.M. (1984) 'General Anatomy of the Gills', in W.S. Hoar and D.J.
    Randall (eds), *Fish Physiology* vol X, Academic Press, Orlando, pp. 1-72
Ishimatsu, A., Johansen, K. and Nilsson, S. (1986) 'Autonomic Nervous Control of
    the Circulatory System in the Airbreathing Fish *Channa argus*', *Comparative
    Biochemistry and Physiology*, in press
Krawkow, N.P. (1913) 'Über die Wirkung von Giften auf die Gefässe isolierter
    Fischkiemen', *Pflügers Archiv für gesamte Physiologie der Menschen und Tiere,
    151*, 583-603
Laurent, P. (1984) 'Gill Internal Morphology', in W.S. Hoar and D.J. Randall
    (eds), *Fish Physiology*, vol. X, Academic Press, Orlando, pp. 73-183
Laurent, P. and Dunel, S. (1976) 'Functional Organization of the Teleost Gill. I.
    Blood Pathways', *Acta Zoologica (Stockholm), 57*, 189-209
Laurent, P. and Dunel, S. (1980) 'Morphology of Gill Epithelia in Fish', *American
    Journal of Physiology, 238*, R147-9
Nakano, T. and Tomlinson, N. (1967) 'Catecholamines and Carbohydrate
    Concentrations in Rainbow Trout (*Salmo gairdneri*) in Relation to Physical
    Disturbance', *Journal of the Fisheries Research Board of Canada, 24*, 1701-15
Nilsson, S. (1973) 'On the Autonomic Nervous Control of Organs in Teleost
    Fishes', in L. Bolis, K. Schmidt-Nielsen and S.H.P. Maddrell (eds),
    Comparative Physiology, North-Holland, Amsterdam, pp. 325-31
Nilsson, S. (1983) *Autonomic Nerve Function in the Vertebrates*, Springer-Verlag,
    Berlin
Nilsson, S. (1984a) 'Innervation and Pharmacology of the Gills', in: W.S. Hoar and
    D.J. Randall (eds), *Fish Physiology*, vol. X, Academic Press, Orlando, pp.
    185-227
Nilsson, S. (1984b) 'Review: Adrenergic Control Systems in Fish', *Marine Biology
    Letters, 5*, 127-46
Nilsson, S. (1985) 'Filamental Position in Fish Gills Is Influenced by a Smooth
    Muscle Innervated by Adrenergic Nerves', *Journal of Experimental Biology,
    118*, 433-437
Nilsson, S. and Pettersson, K. (1981) 'Sympathetic Nervous Control of Blood Flow
    in the Gills of the Atlantic Cod, *Gadus morhua*', *Journal of Comparative
    Physiology, 144*, 157-63
Pärt, P., Kiessling, A. and Ring, O. (1982) 'Adrenaline Increases Vascular
    Resistance in Perfused Rainbow Trout (*Salmo gairdneri* Rich.) Gills',
    Comparative Biochemistry and Physiology, 72C, 107-8
Pärt, P., Tuurala, H., Nikinmaa, M. and Kiessling, A. (1984) 'Evidence for a
    Non-respiratory Intralamellar Shunt in Perfused Rainbow Trout Gills',
    Comparative Biochemistry and Physiology, 79A, 29-34
Pettersson, K. (1983) 'Adrenergic Control of Oxygen Transfer in Perfused Gills of
    the Cod, *Gadus morhua*', *Journal of Experimental Biology, 102*, 327-35
Pettersson, K. and Johansen, K. (1982) 'Hypoxic Vasoconstriction and the Effects
    of Adrenaline on Gas Exchange Efficiency in Fish Gills', *Journal of
    Experimental Biology, 97*, 263-72
Pettersson, K. and Nilsson, S. (1979) 'Nervous Control of the Branchial Vascular
    Resistance of the Atlantic Cod, *Gadus morhua*', *Journal of Comparative
    Physiology, 129*, 179-83
Primmett, D.R.N., Randall, D.J., Mazeaud, M. and Boutilier, R.G. (1986) 'The
    Role of Catecholamines in Erythrocyte pH Regulation and Oxygen Transport
    in Rainbow Trout (*Salmo gairdneri*) during Exercise', *Journal of Experimental
    Biology*, in press
Randall, D.J. and Daxboeck, C. (1984) 'Oxygen and Carbon Dioxide Transfer
    across Fish Gills', in W.S. Hoar and D.J. Randall (eds), *Fish Physiology*, vol.
    X, Academic Press, Orlando, pp. 263-314

Randall, D.J., Baumgarten, D. and Maluysz, M. (1972) 'The Relationship between Gas and Ion Transfer across the Gills of Fishes', *Comparative Biochemistry and Physiology, 41A*, 629-37

Rowing, C.G.M. (1981) 'Interrelationships between Arteries, Veins and Lymphatics in the Head Region of the Eel, *Anguilla anguilla L.*', *Acta Zoologica (Stockholm), 62*, 159-70

Vogel, W. (1978) 'Arterio-venous Anastomoses in the Afferent Region of Trout Gill Filaments (*Salmo gairdneri* Richardson, Teleostei)', *Zoomorphologie, 90*, 205-12

Vogel, W.O.P. (1981) 'Struktur und Organisationsprinzip im Gefässystem der Knochenfische', *Gegenbaurs Morphologische Jahrbuch Leipzig, 127*, 772-84

Vogel, W., Vogel, V. and Schlote, W. (1974) 'Ultrastructural Study of Arterio-venous Anastomoses in Gill Filaments of *Tilapia mossambica*', *Cell and Tissue Research, 155*, 491-512

Wahlqvist, I. (1980) 'Effects of Catecholamines on Isolated Systemic and Branchial Vascular Beds of the Cod, *Gadus morhua*', *Journal of Comparative Physiology, 137*, 139-43

Wahlqvist, I. (1981) 'Branchial Vascular Effects of Catecholamines Released from the Head Kidney of the Atlantic Cod, *Gadus morhua*', *Molecular Physiology, 1*, 235-41

Wahlqvist, I. and Nilsson, S. (1980) 'Adrenergic Control of the Cardio-Vascular System of the Atlantic Cod, *Gadus morhua*, during "stress"', *Journal of Comparative Physiology, 137*, 145-50

Wood, C.M. (1974) 'A Critical Examination of the Physical and Adrenergic Factors Affecting Blood Flow through the Gills of the Rainbow Trout', *Journal of Experimental Biology, 60*, 241-65

Wood, C.M. and Randall, D.J. (1973) 'The Influence of Swimming Activity on Sodium Balance in the Rainbow Trout (*Salmo gairdneri*)', *Journal of Comparative Physiology, 82*, 207-33

6 EXERCISE

Pat J. Butler

The level of exercise performed by fishes, even within one species, is extremely varied. In a loch, the mean hourly swimming speed of brown trout, *Salmo trutta*, did not exceed 0.2 body lengths per second (bls^{-1}) except in March when it peaked at $0.28 bls^{-1}$ (Holliday, Tytler and Young 1972/73). When migrating up river, however, sea trout (the migratory form of brown trout) swim at between 0.5 and $1 bls^{-1}$ *over the ground* for periods of 3-4h at a time (D. Solomon, personal communication). As water velocity may be anything from 0.3 to $1.5 ms^{-1}$, these fish are clearly exercising at high sustainable levels, even if only to maintain position in midstream. Short-duration bursts of activity may be used when negotiating rapid currents during spawning migration and when attempting to capture prey, or when attempting to escape a predator.

Swimming behaviour can be divided into three major categories: sustained, prolonged and burst swimming (Beamish 1978). Sustained swimming can be maintained for long periods ($>$ 200 min) whereas burst swimming can be maintained only for short periods ($<$ 20s). Between these two extremes is prolonged swimming which eventually ends in fatigue. Unless it is possible to track a fish in the wild or swim it in the laboratory for longer than 3h, it is not always possible to distinguish between sustained and prolonged swimming. In the present review, therefore, the distinction will be made between extended periods of swimming (sustained/prolonged) and short-term (burst) swimming, with emphasis placed on the former.

Locomotor Muscles

Although there are basically two (but sometimes three) different types of muscle fibre in fishes, they are not always associated with the different forms of swimming behaviour in a simple way (Johnston 1981). The bulk of muscle consists of so-called 'white' fibres. Typically these have low concentrations of myoglobin, few capillaries and mitochondria and depend primarily on anaerobic glycogenolysis for their energy supply. In round-bodied fish, the so-called 'red' fibres form a superficial sheet of muscle in the midline on each side of the body (Figure 6.1a) which runs from behind the head to the tail fin. In general these have numerous capillaries and have high concentrations of myoglobin, mitochondria and aerobic enzymes. They also contain a greater quantity of glycogen as well as of lipids than the white fibres. The red fibres comprise from 0.5 to 30 per cent of the myotomal muscle, and those fish with a more active mode of life have a higher proportion (Greer-Walker and Pull 1975). In rainbow trout, *Salmo gairdneri*, there is a wide range in the diameter of the white fibres which at the lower end overlaps that of the red fibres. It is often stated that red fibres are intermingled with white fibres in the deeper portion of the myotome of this fish, giving rise to a mosaic arrangement (see, for example, Randall and Daxboeck 1982). Johnston (1981), however, is of the opinion that the range of fibre size in the white muscle represents different stages in growth rather than distinct fibre types.

In tuna, e.g. *Katsuwonus pelamis*, the red muscle may extend internally to the centre of the animal (Figure 6.1b) and is associated with a countercurrent heat exchanger which enables the temperature of the red muscle to remain elevated over a wide range of ambient temperatures. The functional significance of this is unknown but it has been suggested that it allows a faster removal of lactic acid and a faster diffusion of oxygen from blood to the mitochondria (Carey 1982). A further suggestion is that it allows the internal red muscle fibres to contract at the same rate as the

Figure 6.1: Diagrammatic Representation of Transverse Section through the Trunk of Different Fishes to Show the Distribution of Fibre Types in the Myotomal Muscle. (a) Brook trout, *Salvelinus fontinalis*, (b) skipjack tuna, *Katsuwonus pelamis*, (c) common carp, *Cyprinus carpio*. (After Johnston 1981.)

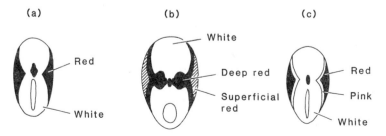

superficial red and white fibres so that the whole of the muscle system can contribute to power generation during burst swimming (Johnston and Brill 1984). In some fish a group of 'pink' fibres are situated between the red and white muscle layers (Figure 6.1c). In the mirror carp, *Cyprinus carpio*, these fibres comprise around 10 per cent of the myotomal muscle and are slightly more numerous than red fibres. Their aerobic capacity is intermediate to those of red and white muscle.

Electromyograph (EMG) recordings have shown that in elasmobranchs, dipnoans and some teleosts, e.g. the Pacific herring, *Clupea pallesi*, only red fibres are active at sustainable swimming speeds. At higher speeds, when the white fibres are recruited, the fish rapidly accumulates lactic acid and soon fatigues. However, in most teleosts, some white fibres are recruited during sustained swimming (Figure 6.2) and there is an initial increase in whole body lactate in small rainbow trout swimming at $3.5\,\mathrm{bl\,s^{-1}}$, but this begins to decrease after 10 min and is similar to the resting value after 24 h of swimming (Wokoma and Johnston 1981). These authors have raised the possibility that the lactate produced by the white muscle is a major substrate for aerobic metabolism in the red fibres. Thus, in the majority of teleosts, the white muscle fibres are not used exclusively during short-term

Figure 6.2: Diagram of Brook Trout, *Salvelinus fontinalis*, Showing Points of Insertion of EMG Electrodes and Traces of EMG Activity Recorded from these Two Sites during Extended Periods of Swimming at 1 and 3 Body Lengths per second ($\mathrm{bl\,s^{-1}}$). Only the red muscle is active at the slower speed but both muscle types are active at the higher speed. (After Johnston and Moon 1980b.)

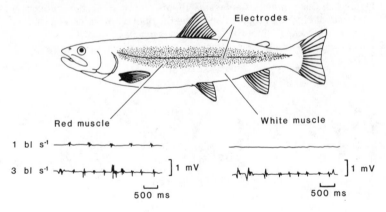

burst swimming, which would soon lead to exhaustion. Whether or not white muscle fibres are active during sustained swimming appears to be related to the type of innervation they receive (Johnston 1981). In elasmobranchs and certain teleosts, each white fibre is innervated by two separate axons which fuse to form a single end plate located at one end of the fibre (focal innervation). In the majority of teleosts the nerves divide and form a diffuse network so that each white fibre is innervated by several different motor axons and receives many nerve terminals (multiple inner-vation) in a similar manner to the red fibres.

Sustained swimming for several days causes hypertrophy of red muscle fibres in the brook trout, *Salvelinus fontinalis*, and of both red and white fibres in the coalfish, *Pollachius virens* (Johnston and Moon 1980a,b). In neither species, however, is there an increase in the activity of aerobic enzymes in red muscles although there is evidence of enhanced fatty acid catabolism in both red and white muscle fibres. This is somewhat different from the situation in mammals where exercise training does result in increased activity of aerobic enzymes (Holloszy and Booth 1976) but is interestingly similar to the situation in birds just before migration (Dawson, March and Yacoe 1983). The similarity is probably related to the fact that aerobic enzyme activity is already high in red muscle of fishes and in the pectoral muscles of birds.

Swimming Performance (U_{crit})

The swimming performance of a fish depends on a number of intrinsic factors such as its size, shape, mode of swimming, the pro-portion of red muscle and the effectiveness of its respiratory and cardiovascular systems in presenting oxygen to the red muscles. It must not be thought, however, that the cardiorespiratory system is always the limiting factor in the exercise performance of a fish, for it may be able to present oxygen at a much higher rate than that required by the aerobic muscle fibres. For example, Soofiani and Priede (1985) have concluded that in juvenile Atlantic cod (*Gadus morhua*) the cardiorespiratory system has evolved to supply the increase in oxygen demand following feeding rather than that during swimming. Extrinsic factors such as temperature, partial pressure of oxygen and environmental pollutants can also influ-ence a fish's ability to swim.

It is not easy to determine a fish's maximum sustainable swimming speed, so performance is often assessed by determining the *critical swimming speed* (Brett 1964) which is a special category of prolonged swimming. This is abbreviated to U_{crit} and is measured by swimming fish in a water channel at successively increasing speeds (say $10\,cm\,s^{-1}$) each lasting for a definite period of time (e.g. 30 min) until the fish fatigues.

$$U_{crit} = U_H + \left[(U_F - U_H) \times \frac{t_F}{t_I} \right]$$

where U_H is the highest velocity maintained for the set time, U_F is the final velocity, t_F is the time the fish swam at the final velocity, and t_I is the set time between velocity increments. If the cross-sectional area of the fish is greater than 10 per cent of the area of the water channel, then a correction should be made for partial blocking of the tube (see Jones, Kiceniuk and Bamford 1974). One problem associated with swimming fish in a water channel is that they are usually unable to use the burst-and-coast form of swimming, which, for fish such as Atlantic cod, may be energetically advantageous at some velocities (Videler and Weihs 1982).

Swimming speed is not always given in absolute terms (i.e. $cm\,s^{-1}$) but rather in terms of the number of body lengths per second. This is presumably to allow comparison between fish of different lengths. Unfortunately, such relative performance at sustained (and critical) speeds is usually higher in smaller individuals of a species, particularly at temperatures close to the physiological optimum (Beamish 1978). It is important, therefore, that whenever possible the length of the fish is stated, whether speed is given in relative or absolute terms.

Respiratory and Circulatory Adjustments

The metabolic, respiratory and cardiovascular adjustments that occur during swimming in fish were reviewed by Jones and Randall (1978). There have only been two comprehensive studies on the topic. Kiceniuk and Jones (1977) swam rainbow trout at 10°C in a Brett-type water channel and related a number of variables to swimming speed in terms of percentage U_{crit}. Piiper, Meyer, Worth and Willmer (1977) took measurements from dog-

fish, *Scyliorhinus stellaris*, at 17°C, that were swimming spontaneously at approximately 0.3bl s^{-1} ($1 \text{bl} \equiv 85 \text{cm}$).

Oxygen uptake increases exponentially with increased swimming speed, and for the rainbow trout this reaches approximately
eight times the resting value at U_{crit} (Figure 6.3). Ventilation
volume increases by approximately the same proportion so that the
percentage of oxygen removed from the water flowing over the
gills (utilisation) remains virtually constant. In dogfish oxygen
uptake increases by 75 per cent whereas ventilation volume is
three times the resting value indicating that utilisation of inspired
oxygen decreases appreciably.

The forward movement of the fish augments the action of the
buccal and opercular pumps, and at speeds of $50\text{-}80 \text{cm s}^{-1}$ many
fishes, e.g. the Atlantic mackerel, *Scomber scombrus*, stop making
rhythmic breathing movements and rely on ram ventilation of the
gills. It is sometimes stated that all scombrid fishes are obligate ram
ventilators, unable to ventilate their gills adequately when not
swimming. This may well be the case for tuna (Jones, Brill and
Mense 1986). There are, however, no differences in blood gases
and pH in the dorsal aorta between swimming and non-swimming
mackerel (Holeton, Pawson and Shelton 1982), which, for this
species at least, dispels this long-standing hypothesis. None the
less, there appears to be an energetic advantage in transferring the
work of ventilation from the branchial to the locomotory muscles
during ram ventilation (Freadman 1979; Steffensen 1985). Indeed,
a switch to ram ventilation at higher swimming speeds is one
explanation for the strange finding that when partial pressure of
oxygen (P_{O_2}) in the environment is low, oxygen uptake of the carp
is lower when swimming at 11.3m s^{-1} than when swimming at
5.6m s^{-1} (Christiansen, Lomholt and Johansen 1982). The transition to ram ventilation may be initiated by mechanoreceptors
detecting water pressure or velocity (see Butler and Metcalfe
1983), but in rainbow trout and the shark sucker, *Echeneis
naucrates*, the water velocity at which transition occurs and the gape
of the mouth are affected by the P_{O_2} of the water (Steffensen 1985).

Whichever form of ventilation (active or ram) is used during
swimming, it is adequate to maintain partial pressure and content
of oxygen in arterial blood (P_aO_2 and C_aO_2 respectively) at or very
close to the resting value in both trout and mackerel. In dogfish,
however, C_aO_2 is approximately 75 per cent of the resting value
during spontaneous swimming although P_aO_2 is only slightly

Figure 6.3: Changes in Oxygen Uptake (V_{O_2} ——), Arterial-mixed Venous Oxygen Content (■), Cardiac Stroke Volume (▲), Heart Rate (△), Systolic (●) and Diastolic (○) Blood pressure in the Dorsal (DA) and Ventral (VA) Aortae of the Rainbow Trout, *Salmo gairdneri*, at Different Swimming Speeds in Water at 10°C. Maximum sustainable swimming speed is designated as U_{crit} and lower speeds are given as percentage of U_{crit}. Points are mean values from nine observations on six trout and vertical lines are ± SE of mean. (From Butler 1985.)

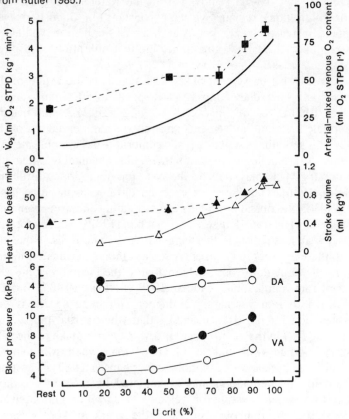

reduced. The arterio-mixed venous difference in oxygen content ($C_aO_2 - C_vO_2$) is maintained in exercising dogfish as C_vO_2 declines by approximately the same amount at C_aO_2. In the more active lemon shark, *Negaprion brevirostris*, however, there is little oxygen in the venous blood at rest and $C_aO_2 - C_vO_2$ is increased during exercise as a result of rises in P_aO_2 and C_aO_2 (Bushnell, Lutz, Steffenson, Oikari and Gruber 1982).

As far as the cardiovascular system is concerned, there is a three times increase in cardiac output in rainbow trout swimming close to U_{crit} and that is achieved predominantly as a result of a 2.2 times increase in cardiac stroke volume. In other words there is little (36 per cent) increase in heart rate. Swimming performance and, by deduction, cardiac performance is not affected by ablation of the coronary artery in trout (Daxboeck 1982). This is because arterial supply to the myocardium is maintained by small arterioles in the adventitia of the ventral aorta and bulbus which appear following the surgery.

Accompanying the increased cardiac output in rainbow trout, there are substantial rises in mean ventral aortic blood pressure and pulse pressure, but much smaller changes occur in the dorsal aorta (Figure 6.3). As such, there is little change in vascular resistance in the branchial blood vessels but a significant (2.5 times) reduction in resistance in the systemic circuit. Not surprisingly it is the red muscle fibres which experience the greatest increase in perfusion during sustained exercise (Randall and Daxboeck 1982), and at 80% U_{crit}, blood flow to these fibres is approximately 14 times the resting value, whereas in the white muscle fibres it is 9 per cent resting. So it is in the red fibres that the greatest vasodilatation occurs, and calculations by Randall and Daxboeck (1982) indicate that 93 per cent of the increased oxygen uptake by the animal as a whole is used by the working muscles. On this basis, these muscles extract 96 per cent of the oxygen in the blood perfusing them, which explains the low mixed venous oxygen content ($C_{\bar{v}}O_2$) in trout swimming at a velocity close to U_{crit}. In fact the 2.5 times increase in ($C_aO_2 - C_{\bar{v}}O_2$) above the resting level makes a substantial contribution to the overall increase in oxygen delivery to the tissues during swimming in trout.

In dogfish, there is a 70 per cent increase in cardiac output during spontaneous swimming and this is very largely the result of an increase in cardiac stroke volume. Heart rate increases by a mere 7 per cent. There is little change in ventral or dorsal aortic blood pressures and as such there is no significant change in systemic vascular resistance and only a slight (43 per cent) fall in resistance in the branchial vascular bed.

The short-finned eel, *Anguilla australis*, is different in its response to exercise from most teleosts that have been studied so far. There is no increase in heart rate or cardiac output, but there is substantial constriction of the branchial blood vessels causing an increase in ventral aortic blood pressure but little change in dorsal aortic pressure (Davie and Forster 1980). The significance of these cardiovascular changes is unclear, but it is possible that these fish cease to swim in response to elevated pressure in the ventral aorta so as to prevent branchial oedema (cf. Satchell 1978).

With the above exception, the relatively small increase in heart rate, compared with the rise in cardiac output, during swimming appears to be common to all fish studied so far. For example, Atlantic cod show a 12 per cent increase above the resting value when swimming close to U_{crit} (Butler 1985). When accelerating from rest or from one level of exercise to another, this animal, together with ling cod, *Ophiodon elongatus*, short-finned eel and rainbow trout (Stevens, Bennion, Randall and Shelton 1972; Priede 1974; Davie and Forster 1980; Butler 1985) exhibit a transient bradycardia. The significance of this is not clear as it occurs in animals swimming spontaneously as well as in those in water channels. It is abolished by injection of atropine, indicating that it is mediated by a rise in vagal parasympathetic tone.

Cardiovascular Control

A reduction in vagal parasympathetic tone may be the cause of the sustained tachycardia during exercise at low environmental temperature but at higher temperatures the vagus nerve does not appear to be of great importance in controlling heart rate during exercise, in rainbow trout at least (Priede 1974). Jones and Randall (1978) conclude that the slight tachycardia of exercise results from the combined effect of increased sympathetic tone (not in elasmobranchs, which have a sparse sympathetic innervation of the heart) and decreased parasympathetic tone together with the effect of elevated levels of circulating catecholamines. Neurohumoral mechanisms are also thought by these authors to be more important than passive factors in causing the increased cardiac stroke volume during exercise.

The presence and control of systemic vascular tone in resting fish differs from species to species. In resting rainbow trout and Atlantic cod, systemic vascular tone results solely from sympathetic nervous activity acting via α-adrenoreceptors (Smith 1978; Smith, Nilsson, Wahlqvist and Eriksson 1985). Recent work indicates that any selective vasoconstriction that occurs during swimming in Atlantic cod is also the exclusive result of sympathetic nervous activity (M. Axelsson and S. Nilsson, in preparation). There appears to be no systemic vascular tone in resting, short-finned eels, although both nervous and humoral mechanisms could cause vasoconstriction during exercise (Hipkins 1985; Hipkins, Smith and Evans 1986). The systemic circulation of the spiny dogfish, *Squalus acanthias*, on the other hand, is unresponsive to stimulation of the CNS but responds to infused catecholamines (Opdyke, McGreehan, Messing and Opdyke 1972). This indicates that only humoral mechanisms are going to be involved in any constriction of the systemic circulation during exercise in this animal. The cause of the massive vasodilation in the red muscle fibres is unclear. Randall and Daxboeck (1982) suggest that 'active hyperaemia' is the major factor involved.

The fact that there is usually little or no change in resistance in the blood vessels of the gills during exercise (Kiceniuk and Jones 1977; Piiper *et al.* 1977) might suggest that there is normally little or no modification of branchial vasomotor tone. However, the ability of the gills to transfer oxygen increases by five to six times during exercise in the rainbow trout (Jones and Randall 1978) and this is indicative of an increase in the area available for gas exchange, a decrease in the diffusion distance between the blood and water and/or an increase in the permeability of the gills to oxygen. It is worth noting, in passing, that these changes at the gills during exercise not only have a beneficial influence on gas exchange; they may also have an adverse effect on ionic and osmotic balance. In rainbow trout, there is an increase in water and ion movements through the gills during exercise (Wood and Randall 1973), and in the Atlantic mackerel, plasma concentrations of Na^+, K^+ and Cl^- rise with increasing swimming speed, especially after transition to ram ventilation (Boutilier, Aughton and Shelton 1984).

Not all of the blood leaving the secondary lamellae of the gills in teleosts and elasmobranchs enters the dorsal aorta to supply the body (see Butler and Metcalfe 1983), and in the eel, catfish *Ictalurus punctatus* and two species of dogfish, *Squalus acanthias* and *Centrophorus scalpratus*, not all of the blood leaving the heart perfuses the secondary lamellae (Steen and Kruyse 1964; Holbert, Boland and Olson *et al.* 1979; Cooke 1980; Olson and Kent 1980; De Vries and De Jager 1984). Thus, by nervous and/or humoral mechanisms it may be possible for fishes to control the proportion of efferent blood that perfuses the body and, in some species, also to control the proportion of afferent blood that perfuses the secondary lamellae (Nilsson 1984). In resting rainbow trout in well-aerated water, only 60 per cent of the secondary lamellae are perfused with blood (Booth 1978) and even in those lamellae that are perfused, up to 20 per cent of the blood may be in basal channels which are probably non-respiratory because of their greater distance from the water than the other lamellar blood spaces (Tuurala, Pärt, Nikinmaa and Soivio 1984). Adrenaline causes interlamellar recruitment *in vivo* (Booth 1979; Holbert *et al.* 1979), and an increase in perfusion pressure causes inter- and intralamellar recruitment *in vitro* (Farrell, Daxboeck and Randall 1979; Farrell, Sobin, Randall and Crosby 1980). Intralamellar recruitment occurs as a result of a greater proportion of blood perfusing central lamellar spaces, i.e. there is a move away from the basal channels. There may also, therefore, be a thinning of the epithelium of the lamellae thus reducing the diffusion distance between blood and water. It has been demonstrated *in vitro* that adrenaline increases the permeability of the gills of freshwater-adapted rainbow trout to butanol and water, and it has been stated that it has the same effect on oxygen (Isaia 1984), although direct experimental evidence is lacking.

Thus, there are a number of ways in which oxygen transport across the gills and to the body could be affected during exercise. A direct sympathetic nervous influence may control the proportion of efferent blood perfusing the body in the Atlantic cod (Nilsson and Pettersson 1981) whereas, as already indicated, circulating catecholamines could influence either directly, or indirectly via a rise in blood pressure, the effective surface area of the gills for gas exchange. In the dogfish, *S. canicula*, there is no direct nervous innervation of the branchial blood vessels (Metcalfe and Butler 1984), so any control of the pattern of blood flow through the gills during exercise in this fish would have to be humoral or mechanical in nature.

Plasma Catecholamines and Exercise

It is often stated that an increase in circulating catecholamines occurs during exercise (Randall 1982) which could, therefore, effect a number of the changes outlined above and augment gas exchange across the gills and thus oxygen delivery to the muscles.

In a more recent review, however, Wood and Perry (1985), while attributing many of the compensatory responses that occur during exercise to circulating catecholamines, admit that their evidence is 'circumstantial rather than direct' because the levels of plasma catecholamines have rarely been measured during 'experimental treatments'.

The evidence that plasma catecholamines increase during exercise is based largely on a paper by Nakano and Tomlinson (1967), who found that in rainbow trout adrenaline increased 26-fold to 7 $\times$ 10^{-7}M and noradrenaline increased 8-fold to 1.5×10^{-7}M. However, these large changes were obtained by 'repeatedly grasping the fish by the tail ... thus forcing them to move vigorously about the tank'. These authors properly refer to this as 'physical disturbance' whereas Opdyke, Carroll and Keller (1982), using a similar technique to that of Nakano and Tomlinson to induce activity, refer to catecholamine release during *exercise* in the dogfish. In a recent study (Butler, Metcalfe and Ginley, 1986), plasma catecholamines have been measured in resting fish and during spontaneous activity (dogfish) or during sustained and exhaustive swimming in a water channel (rainbow trout). No electric grid or prodding was used in this part of the study. In both cases, however, the fishes were induced into burst swimming for 2-3 min by repeatedly touching their tails.

During spontaneous swimming in dogfish there is a 2- to 3-fold increase in plasma catecholamines, which is statistically significant for adrenaline only. During tail grasping there is a 7- to 16-fold increase to approximately 10^{-7}M, which is significant in both cases (Figure 6.4a). There is no increase in either adrenaline or noradrenaline in the rainbow trout during sustained swimming; in fact in these experiments the level of noradrenaline actually decreased (Figure 6.4b). Even when swimming to apparent exhaustion in a water channel, the levels obtained are not high at 1.2×10^{-8}M for adrenaline and 2.3×10^{-8}M for noradrenaline. Only during repeated burst swimming do the levels approach those reported by Nakano and Tomlinson (1967), i.e. approximately 10^{-7}M. Similar studies on Atlantic cod have also demonstrated no changes in plasma catecholamines above the resting values of approximately 7 $\times$ 10^{-9}M during sustained swimming (M. Axelsson and S. Nilsson in preparation) and levels of approximately 4×10^{-8}M when swimming to apparent exhaustion (P.J. Butler, J.D. Metcalfe, M. Axelsson and S. Nilsson in preparation).

Figure 6.4: Mean Values (± SE of Mean) of Plasma Noradrenaline and Adrenaline in Dogfish and Trout at Rest, while Swimming (Spontaneously in the Dogfish and in a Water Channel in the Trout) and in Response to Repeated Burst Swimming for 2-3 min. Number of animals used is given in parentheses beneath each histogram. (Data from P.J. Butler, J.D. Metcalfe and S.A. Ginley, in preparation.)

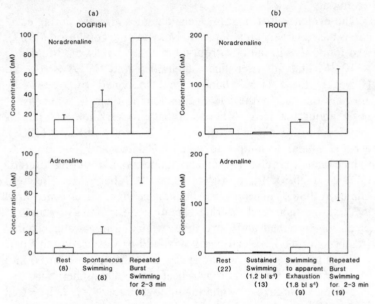

The physiological effects of adrenaline on gas exchange across the gills and on the characteristics of the oxygen equilibrium curve have been demonstrated using concentrations of 10^{-5} to 10^{-7}M (Wood, McMahon and McDonald 1978; Nikinmaa 1982; Pettersson and Johansen 1982; Pettersson 1983), which are higher than those measured in rainbow trout and Atlantic cod swimming to exhaustion but close in some cases to those measured in rainbow trout during tail grasping. It would seem, therefore, that plasma catecholamines are not going to reach a level of overall physiological significance as far as oxygen delivery to the tissues is concerned during swimming itself, even if the fish is temporarily exhausted. However, if repeated burst swimming occurs, perhaps as the result of some sort of physical disturbance (which in natural conditions could include being chased by a predator), then plasma catecholamines could well reach an effective level. This would be the classical 'fight or flight' response. The extreme nature of prolonged bouts of burst swimming is indicated by the fact that approximately 40 per cent of rainbow trout die within 4-8 h following such activity, possibly as the result of intracellular acidosis (Wood, Turner and Graham 1983).

The notion that plasma catecholamines are not important in oxygen delivery to the muscles during exercise is supported by results from current experiments on Atlantic cod. It is known that environmental hypoxia and damage to the gills can affect swimming performance in fish (Jones 1971; Duthie and Hughes 1982; Bushnell, Steffensen and Johansen 1984). So, if plasma catecholamines do play an important role in providing the increased oxygen supply to the active muscles during exercise, then preventing their release should impair swimming performance. Evidence so far indicates that, although sectioning nerves to the head kidney prevents the increase in plasma catecholamines in cod swimming to exhaustion, it does not affect the fishes' swimming performance (P.J. Butler, J.D. Metcalfe, M. Axelsson and S. Nilsson, in preparation). It is necessary to look elsewhere for the factors that control and co-ordinate the cardiovascular responses that ensure an adequate supply of oxygen to the working muscles during exercise in fish.

References

Beamish, F.W.H. (1978) 'Swimming Capacity', in W.S. Hoar and D.J. Randall (eds.), *Fish Physiology, Vol. VII*, Academic Press, New York, pp. 101-87

Booth, J.H. (1978) 'The Distribution of Blood Flow in the Gills of Fish: Application of a New Technique to Rainbow Trout (*Salmo gairdneri*)', *Journal of Experimental Biology, 73*, 119-29

Booth, J.H. (1979) 'The Effects of Oxygen Supply, Epinephrine, and Acetylcholine on the distribution of Blood Flow in Trout Gills', *Journal of Experimental Biology, 83*, 31-9

Boutilier, R.G., Aughton, P. and Shelton, G. (1984) 'O_2 and CO_2 Transport in Relation to Ventilation in the Atlantic Mackerel, *Scomber scombrus*', *Canadian Journal of Zoology, 62*, 546-54

Brett, J.R. (1964) 'The Respiratory Metabolism and Swimming Performance of Young Sockeye Salmon', *Journal of the Fisheries Research Board of Canada, 21*, 1183-226

Bushnell, P.G., Lutz, P.L., Steffensen, J.F., Oikari, A. and Gruber, S.H. (1982) 'Increases in Arterial Blood Oxygen during Exercise in the Lemon Shark (*Negaprion brevirostris*)', *Journal of Comparative Physiology, 147*, 41-7

Bushnell, P.G., Steffensen, J.F. and Johansen, K. (1984) 'Oxygen Consumption and Swimming Performance in Hypoxia-acclimated Rainbow Trout *Salmo gairdneri*', *Journal of Experimental Biology, 113*, 225-35

Butler, P.J. (1985) 'Exercise in Non-mammalian Vertebrates: a Review', *Journal of the Royal Society of Medicine, 78*, 739-47

Butler, P.J. and Metcalfe, J.D. (1983) 'Control of Respiration and Circulation', in
J.C. Rankin, T.J. Pitcher and R.T. Duggan (eds), *Control Processes in Fish
Physiology*, Croom Helm, London, pp. 41-65

Butler, P.J., Metcalfe, J.D. and Ginley, S.A. (1986) 'Plasma Catecholamines in the
Lesser Spotted Dogfish and Rainbow Trout at Rest and During Different
Levels of Exercise', *Journal of Experimental Biology*, in press

Carey, F.G. (1982) 'Warm Fish', in C.R. Taylor, K. Johansen and L. Bolis (eds), *A
Companion to Animal Physiology*, Cambridge University Press, Cambridge,
pp. 216-33

Christiansen, B., Lomholt, J.P. and Johansen, K. (1982) 'Oxygen Uptake of Carp,
Cyprinus carpio, Swimming in Normoxic and Hypoxic Water', *Environmental
Biology of Fish*, 7, 291-6

Cooke, I.R.C. (1980) 'Functional Aspects of the Morphology and Vascular
Anatomy of the Gills of the Endeavour Dogfish, *Centrophorus scalpratus*
(McCulloch) (Elasmobranchii: Squalidae)', *Zoomorphologie*, 94, 167-83

Davie, P.S. and Forster, M.E. (1980) 'Cardiovascular Responses to Swimming in
Eels', *Comparative Biochemistry and Physiology*, 67, 367-73

Dawson, W.R., March, R.L. and Yacoe, M.E. (1983) 'Metabolic Adjustments of
Small Passerine Birds for Migration and Cold', *American Journal of Physiology*,
245 (*Regulatory Integrative Comparative Physiology, 14*), R755-67

Daxboeck, C. (1982) 'Effect of Coronary Artery Ablation on Exercise
Performance in *Salmo gairdneri*', *Canadian Journal of Zoology*, 60, 375-81

De Vries, R. and De Jager, S. (1984) 'The Gill of the Spiny Dogfish, *Squalus
acanthias: Respiratory and Nonrespiratory Function*', *The American Journal of
Anatomy*, 169, 1-29

Duthie, G.G. and Hughes, G.M. (1982) 'Some Effects of Gill Damage on the
Swimming Performance of Rainbow Trout (*Salmo gairdneri*), *Journal of
Physiology*, 327, 21-2P

Farrell, A.P., Daxboeck, C. and Randall, D.J. (1979) 'The Effect of Input Pressure
and Flow on the Pattern and Resistance to Flow in the Isolated Perfused Gill of
a Teleost Fish', *Journal of Comparative Physiology*, 133, 233-40

Farrell, A.P., Sobin, S.S., Randall, D.J. and Crosby, S. (1980) 'Intralamellar Blood
Flow Patterns in Fish Gills', *American Journal of Physiology*, 239 (*Regulatory
Integrative Comparative Physiology, 8*), R428-36

Freadman, M.A. (1979) 'Swimming Energetics of Striped Bass (*Morone saxatilis*)
and Bluefish (*Pomatomus saltatrix*): Gill Ventilation and Swimming
Metabolism', *Journal of Experimental Biology*, 83, 217-30

Greer-Walker, M. and Pull, G.A. (1975) 'A Survey of Red and White Muscle in
Marine Fish', *Journal of Fish Biology*, 7, 295-300

Hipkins, S.F. (1985) 'Adrenergic Responses of the Cardiovascular System of the
Eel, *Anguilla australis, in vivo*', *Journal of Experimental Zoology*, 235, 7-20

Hipkins, S.F., Smith, D.G. and Evans, B.K. (1986) 'Lack of Adrenergic Control of
Dorsal Aortic Blood Pressure in the Resting Eel, *Anguilla australis*', *Journal of
Experimental Zoology*, in press

Holbert, P.W., Boland, E.J. and Olson, K.R. (1979) 'The Effect of Epinephrine
and Acetylcholine on the Distribution of Red Cells within the Gills of the
Channel Catfish (*Ictalurus punctatus*)', *Journal of Experimental Biology*, 79,
135-46

Holeton, G.F., Pawson, M.G. and Shelton, G. (1982) 'Gill Ventilation, Gas
Exchange, and Survival in the Atlantic Mackerel (*Scomber scombrus* L.)',
Canadian Journal of Zoology, 60, 1141-7

Holliday, F.G.T., Tytler, P. and Young, A.H. (1972/73) 'Activity Levels of Trout
(*Salmo trutta*) in Airthrey Loch, Stirling, and Loch Leven, Kinross',
Proceedings of the Royal Society of Edinburgh B, 74, 315-31

Holloszy, J.O. and Booth, F.W. (1976) 'Biochemical Adaptations to Endurance

Exercise in Muscle', *Annual Reviews in Physiology, 1151,* 273-91

Isaia, J. (1984) 'Water and Nonelectrolyte Permeation', in W.S. Hoar and D.J. Randall (eds), *Fish Physiology, vol. X,* Academic Press, New York, pp. 1-38

Johnston, I.A. (1981) 'Structure and Function of Fish Muscles', *Symposia of the Zoological Society, London, 48,* 71-113

Johnston, I.A. and Brill, R. (1984) 'Thermal Dependence of Contractile Properties of Single-skinned Muscle Fibres from Antarctic and Various Warm Water Marine Fishes Including Skipjack Tuna (*Katsuwonus pelamis*) and Kawakawa (*Euthynnus affinis*)', *Journal of Comparative Physiology, 155,* 63-70

Johnston, I.A. and Moon, T.W. (1980a) 'Endurance Exercise Training in the Fast and Slow Muscles of a Teleost Fish (*Pollachius virens*)', *Journal of Comparative Physiology, 135,* 147-56

Johnston, I.A. and Moon, T.W. (1980b) 'Exercise Training in Skeletal Muscle of Brook Trout (*Salvelinus fontinalis*)', *Journal of Experimental Biology, 87,* 177-94

Jones, D.R. (1971) 'The Effect of Hypoxia and Anaemia on the Swimming Performance of Rainbow Trout (*Salmo gairdneri*)', *Journal of Experimental Biology, 55,* 541-51

Jones, D.R. and Randall, D.J. (1978) 'The Respiratory and Circulatory Systems during Exercise', in W.S. Hoar and D.J. Randall (eds), *Fish Physiology,* vol. VII, Academic Press, New York, pp. 425-501

Jones, D.R., Kiceniuk, J.W. and Bamford, O.S. (1974) 'Evaluation of the Swimming Performance of Several Fish Species from the Mackenzie River', *Journal of the Fisheries Research Board of Canada, 31,* 1641-7

Jones, D.R., Brill, R.W. and Mense, D.C. (1986) 'The Influence of Blood Gas Properties on Gas Tensions and pH of Ventral and Dorsal Aortic Blood in Free Swimming Tuna (*Euthynnus affinis*)', *Journal of Experimental Biology, 120,* 201-31

Kiceniuk, J.W. and Jones, D.R. (1977) 'The Oxygen Transport System in Trout (*Salmo gairdneri*) during Sustained Exercise', *Journal of Experimental Biology, 69,* 247-60

Metcalfe, J.D. and Butler, P.J. (1984) 'On the Nervous Regulation of Gill Blood Flow in the Dogfish (*Scyliorhinus canicula*)', *Journal of Experimental Biology, 113,* 253-67

Nakano, T. and Tomlinson, N. (1967) 'Catecholamine and Carbohydrate Concentrations in Rainbow Trout (*Salmo gairdneri*) in Relation to Physical Disturbance', *Journal of the Fisheries Research Board of Canada, 24,* 1701-15

Nikinmaa, M. (1982) 'The Effects of Adrenaline on the Oxygen Transport Properties of *Salmo gairdneri* Blood', *Comparative Biochemistry and Physiology, 71,* 353-6

Nilsson, S. (1984) 'Innervation and Pharmacology of the Gills', in W.S. Hoar and D.J. Randall (eds), *Fish Physiology,* vol. X, Academic Press, New York, pp. 185-227

Nilsson, S. and Pettersson, K. (1981) 'Sympathetic Nervous Control of Blood Flow in the Gills of the Atlantic Cod, *Gadus morhua*', *Journal of Comparative Physiology, 144,* 157-63

Olson, K.R. and Kent, B. (1980) 'The Microvasculature of the Elasmobranch Gill', *Cell and Tissue Research, 209,* 49-63

Opdyke, D.F., McGreehan, J.R., Messing, S. and Opdyke, N.E. (1972) 'Cardiovascular Responses to Spinal Cord Stimulation and Autonomically Active Drugs in *Squalus acanthias*', *Comparative Biochemistry and Physiology, 42,* 611-20

Opdyke, D.F., Carroll, R.G. and Keller, N.E. (1982) 'Catecholamine Release and Blood Pressure Changes Induced by Exercise in Dogfish', *American Journal of Physiology, 242 (Regulatory Integrative Comparative Physiology, 11),* R306-10

Pettersson, K. (1983) 'Adrenergic Control of Oxygen Transfer in Perfused Gills of the Cod, *Gadus morhua*', *Journal of Experimental Biology, 102,* 327-35

Pettersson, K. and Johansen, K. (1982) 'Hypoxic Vasoconstriction and the Effects of Adrenaline on Gas Exchange Efficiency in Fish Gills', *Journal of Experimental Biology, 97,* 263-72

Piiper, J., Meyer, M., Worth, H. and Willmer, H. (1977) 'Respiration and Circulation during Swimming Activity in the Dogfish *Scyliorhinus stellaris*', *Respiration Physiology, 30,* 221-39

Priede, I.G. (1974) 'The Effect of Swimming Activity and Section of the Vagus Nerves on Heart Rate in Rainbow Trout', *Journal of Experimental Biology, 60,* 305-19

Randall, D. (1982) 'The Control of Respiration and Circulation in Fish during Exercise and Hypoxia', *Journal of Experimental Biology, 100,* 275-88

Randall, D.J. and Daxboeck, C. (1982) 'Cardiovascular Changes in the Rainbow Trout (*Salmo gairdneri* Richardson) during Exercise', *Canadian Journal of Zoology, 60,* 1135-40

Satchell, G.H. (1978) 'Type J Receptors in the Gills of Fish', in C. Porter (ed.), *Studies in Neurophysiology,* Cambridge University Press, pp. 131-42

Smith, D.G. (1978) 'Neural Regulation of Blood Pressure in Rainbow Trout (*Salmo gairdneri*)', *Canadian Journal of Zoology, 56,* 1678-83

Smith, D.G., Nilsson, S., Wahlqvist, I. and Eriksson, B-M. (1985) 'Nervous Control of the Blood Pressure in the Atlantic Cod, *Gadus morhua*', *Journal of Experimental Biology, 117,* 335-47

Soofiani, N.M. and Priede, I.G. (1985) 'Aerobic Metabolic Scope and Swimming Performance in Juvenile Cod, *Gadus morhua* L.', *Journal of Fish Biology, 26,* 127-38

Steen, J.B. and Kruysse, A. (1964) 'The Respiratory Function of Teleostean Gills', *Comparative Biochemistry and Physiology, 12,* 127-42

Steffensen, J.F. (1985) 'The Transition between Branchial Pumping and Ram Ventilation in Fishes: Energetic Consequences and Dependence on Water Oxygen Tension', *Journal of Experimental Biology, 114,* 141-50

Stevens, E.D., Bennion, G.R. Randall, D.J. and Shelton, G. (1972) 'Factors Affecting Arterial Pressures and Blood Flow from the Heart in Intact, Unrestrained Lingcod, *Ophiodon elongatus*', *Comparative Biochemistry and Physiology, 43,* 681-95

Tuurala, H., Part, P., Nikinmaa, M. and Soivio, A. (1984) 'The Basal Channels of Secondary Lamellae in *Salmo gairdneri* Gills — a Non-respiratory Shunt', *Comparative Biochemistry and Physiology, 79,* 35-9

Videler, J.J. and Weihs, D. (1982) 'Energetic Advantages of Burst-and-coast Swimming of Fish at High Speeds', *Journal of Experimental Biology, 97,* 169-78

Wokoma, A. and Johnston, I.A. (1981) 'Lactate Production at High Sustainable Cruising Speeds in Rainbow Trout (*Salmo gairdneri* Richardson)', *Journal of Experimental Biology, 90,* 361-4

Wood, C.M. and Perry, S.F. (1985) 'Respiratory, Circulatory, and Metabolic Adjustments to Exercise in Fish', in R. Gilles (ed.), *Circulation, Respiration and Metabolism. Current Comparative Approaches,* Springer-Verlag, Berlin, pp. 2-22

Wood, C.M. and Randall, D.J. (1973) 'The Influence of Swimming Activity on Sodium Balance in the Rainbow Trout (*Salmo gairdneri*)', *Journal of Comparative Physiology, 82,* 207-33

Wood, C.M., McMahon, B.R. and McDonald, D.G. (1978) 'Oxygen Exchange and Vascular Resistance in the Totally Perfused Rainbow Trout', *American Journal of Physiology, 239 (Regulatory Integrative Comparative Physiology, 3),* R201-8

Wood, C.M., Turner, J.D. and Graham, M.S. (1983) 'Why Do Fish Die after Severe Exercise?' *Journal of Fish Biology, 22,* 189-201

7 GASTRO-INTESTINAL PEPTIDES IN FISH

Susanne Holmgren, Ann-Cathrine Jönsson and Björn Holstein

Small polypeptides with a presumed biological regulatory function found in nerves and endocrine mucosal cells of the gut are usually referred to as gastro-intestinal peptides or gastro-intestinal hormones. Both expressions are inadequate, but commonly used. A third often-used expression is 'brain-gut' peptides, indicating the dual location of most of these peptides in the brain and the gut. The term 'regulatory peptides' refers to the involvement of these peptides in a number of physiological functions.

Studies within this field have progressed dramatically over the last decade. This is mainly due to the development of immunological methods enabling scientists to localise and measure the amount of different smaller molecules, especially polypeptides, by the use of specific antibodies. The picture of a complicated system of interacting neurons and endocrine cells in the gut has thus emerged. More and more information is obtained every day, especially in the histological and biochemical fields describing and characterising the structure and distribution of more than 30 different peptides in the gut. The physiology is progressing at a slower rate; experiments are more time-consuming, often difficult to evaluate (one or two out of possibly 20 interacting parameters are studied), and in most cases pharmacological tools are so far lacking.

Most studies of the regulatory peptides deal with the situation in mammals, but there is great interest in the evolution of these peptides. Some peptides seem to be well preserved throughout evolution and it is usual that at least the biologically active site is conserved, whereas other peptides have changed considerably. The available antisera are, with few exceptions, raised against the mammalian peptide sequence, and this may mean that some antisera are unsuitable for work on lower vertebrates, failing to detect even a closely related peptide. An antiserum may on the other hand recognise a peptide not identical with the one that was the original hapten for the antibody. The results of identification with antibodies therefore commonly refer to, for example,

119

bombesin-*like* immunoreactivity (IR) rather than bombesin IR.

In this review we have, however, for reasons of space, chosen to use the incorrect but shorter expression, although it is highly probable that in most cases the fish peptide identified varies more or less in amino acid sequence from its mammalian counterpart.

This chapter attempts to collect available information on the presence of regulatory peptides in endocrine cells and nerves of the fish gut. The possible physiological function of some of these peptides in control of motility and gastric secretion will also be discussed. When possible, examples are given both from elasmobranch and actinopterygian (holostean, teleostean) fish. For comparison, some available results from cyclostomes (*Myxine, Lampetra*) are included.

Endocrine Cells in the Gastro-intestinal Tract of Fish

The endocrine cells of the gastro-intestinal tract are located in the mucosa. 'True' endocrine cells deliver their content to the bloodstream, and the target organ can thus be at any distance from the endocrine cell. The endocrine cells are of two types: open and closed. The open type is a cell that most often has its base at the basal lamina and extends to the lumen with an apical process. The open endocrine cell is considered to be the most primitive of the two (Östberg, Van Noorden, Pearse and Thomas 1976). The mammalian endocrine cells of the gastro-intestinal canal can be divided into at least 14-16 subtypes, based on ultrastructural and immunohistochemical evidence (cf. Solcia, Capella, Buffa, Frigerio, Usellini and Fiocca 1980). Some endocrine cells have, in addition to the apical process, basal processes that reach out and make contact with neighbouring cells, giving them a paracrine (a local control) function.

The endocrine cells of the fish gut seem mainly to be of the open type, i.e. flask-shaped, with the base at the lamina and an apical process reaching the lumen. A number of brain-gut peptides have been localised in endocrine cells. Up to eight different subtypes of endocrine cells have been reported in a teleost fish, *Barbus conchonius* (Rombout and Reinecke 1984), compared with 14-16 in mammals (Solcia *et al.* 1980). A brief summary of brain-gut peptides of endocrine cells in fish is given below (see also Table 7.1).

Table 7.1: The Occurrence of Immunoreactive Nerve Fibres (○) and Endocrine Cells (●) in the Gastro-intestinal Tract of Different Fish Species

	BM	ENK	G/CCK	GLUC	INS	NT	PHI	PP	SOM	SP	VIP
Cyclostomes											
Lampetra fluviatilis (1,2)			●	●	●				●		
Myxine glutinosa (3,4,5,6,7)			●	●	●	●			●		●
Elasmobranchs											
Squalus acanthias (5,6,7,8,9,10)	◐	●	◐	●		●			◐	◐	◐
Holosteans											
Lepisosteus platyrhincus (11)	○				○						○
Teleosts											
Ameiurus nebulosus (12)			●								
Anguilla anguilla (2)		○								●	○
Barbus conchonius (13,14)	◐	●	●		◐	○			◐	◐	◐
Brachydanio rerio (15)				●		●					○
Carassius auratus (16)					●						
Corydoras schultzei (15)		○	●	●					●		○
Cyprinus carpio (12)			●								
Gadus morhua (17,18)	○								○	○	
Gillichtys mirabilis (19)		○				○			○	○	
Gyrinocheilus aymonieri (15)									●		○
Haplochromis sp (15)	◐	●		●				●	●		
Helostoma temmincki (15)		○	●	●	○			●	●		○
Hemigrammus ocellifer (15)		●	●	●							○
Idus idus (15)		●	●	●				●	●		○
Myoxocephalus scorpius = (=*Cottus scorpius*)(5,6,7,17)		○				●			●		●
Pantodon buchholzi (15)			●	●	●						○
Perca fluviatilis (12)		●						●			
Pelmatochromus pulcher (15)		●			●			●	●	●	
Platypoecilus pulcher = (15) (=*platypoecilius variatus*)		●	●			○		●			
Poecilia reticulata (15)		●	●			○					○
Raniceps raninus (5,6)						●					●
Salmo gairdneri (17,20)	◐	●	●						●	◐	○

Numbers within parentheses refer to the following references: 1. Van Noorden and Pearse (1974); 2. Van Noorden and Falkmer (1980); 3. Östberg *et al.* (1976); 4. Falkmer and Östberg (1977); 5. Reinecke *et al.* (1980b); 6. Reinecke *et al.* (1981); 7. Reinecke *et al.* (1984); 8. Holmgren and Nilsson (1983a); 9. El Salhy (1984); 10. Holmgren (1985); 11. Holmgren and Nilsson (1983b); 12. Noaillac-Depeyre and Hollande (1981); 13. Rombout and Taverne-Thiele (1982); 14. Rombout and Reinecke (1984); 15. Langer *et al.* (1979); 16. Reinecke *et al.* (1980a); 17. Thorndyke *et al.* (1984); 18. Jensen and Holmgren (1985); 19. Van Noorden and Patent (1980); 20. Holmgren *et al.* (1982). Legend; BM − bombesin, ENK − enkephalin, G/CCK − gastrin/cholecystokinin, GLUC − glucagon, INS − insulin, NT − neurotensin, PP − pancreatic polypeptide, SOM − somatostatin, SP − substance P, VIP − vasoactive intestinal polypeptide

Bombesin

Bombesin is a tetradecapeptide first isolated from the skin of a frog, and closely related to the mammalian GRP (gastrin releasing peptide), both having a potent gastrin releasing effect in mammals (Anastasi, Erspamer and Bucci 1971; McDonald, Nilsson, Vagne, Ghatei, Bloom and Mutt 1978).

In the elasmobranch, *Squalus acanthias,* the stomach, intestinal and rectal mucosa contains bombesin IR endocrine cells (Figure 7.1A; Holmgren and Nilsson, 1983a; El Salhy 1984). It is possible that the amino acid sequence of elasmobranch bombesin closely resembles that of the tachykinins, as the IR is quenched by 90 per

Figure 7.1: Neuropeptides in Endocrine Cells and Nerves of the Fish Gut. A, Bombesin-like IR in endocrine cells in the mucosa of the rectum of *Squalus acanthias*; calibration bar = 200 μm. B, Endocrine cell in the intestinal mucosa of *Squalus acanthias* showing somatostatin-like IR; calibration bar = 50 μm. C. Nerve fibres in the myenteric plexus of the cardiac stomach from *Gadus morhua* showing substance-P like IR; calibration bar = 100 μm

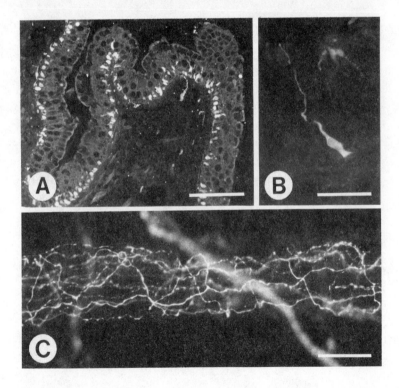

cent by preabsorbtion of bombesin antisera with substance P (Holmgren and Nilsson 1983a).

Among teleosts, endocrine cells containing bombesin IR material are found in the stomach and distal intestine of *Salmo gairdneri* (Holmgren, Vaillant and Dimaline 1982).

Enkephalin

Two endogenous opiate peptides, met- and leu-enkephalin, exerting strong motor effects on the gut via opiate receptors, are present both in the gut and the brain of mammals (Jaffe and Martin 1975). In fish, endocrine cells showing IR to enkephalins have been found in the intestine and rectum of *Squalus acanthias*. In addition alpha- and beta-endorphin, which both terminate with the sequence of met-enkephalin, were present in the posterior intestine and duodenum, respectively (Holmgren and Nilsson 1983a; El Salhy 1984).

Only a few teleost species have been found to possess enkephalin IR cells in the gut (Langer, Van Noorden, Polak and Pearse 1979; Van Noorden and Falkmer 1980; Rombout and Taverne-Thiele 1982; Rombout and Reinecke 1984).

Gastrin, Cholecystokinin and Caerulein

Gastrin, a powerful stimulator of gastric acid secretion, was discovered as early as 1905 by Edkins, but was not sequenced until 1964 (by Gregory and Tracy). Cholecystokinin (CCK)/ pancreozymin with, for example, stimulatory action on gallbladder contractility, was demonstrated by Ivy and Oldberg in 1928, and later sequenced by Mutt and co-workers in 1968-1971 (for references see Sundler, Sjölund and Håkansson 1983). These two peptides and the amphibian peptide caerulein (Anastasi, Erspamer and Endean 1968) share a common C-terminal pentapeptide amide, which is the biologically active part of this peptide family, giving the peptides overlapping physiological effects.

In the cyclostome, *Lampetra fluviatilis*, cells in the intestinal epithelium contain gastrin- and caerulein-IR; the same cells also show IR against glucagon antibodies (Van Noorden and Pearse 1974). *Myxine glutinosa*, the hagfish, also has gastrin/CCK IR cells in the intestine (Östberg *et al.* 1976).

The elasmobranch, *Squalus acanthias*, has scattered gastrin/ CCK IR cells in the intestinal mucosa (Holmgren and Nilsson 1983a; El Salhy 1984). In addition, El Salhy (1984) reports the presence of gastrin/CCK-positive cells in the rectal mucosa. A CCK-specific antiserum showed endocrine cells with the same distribution but with lower density than that shown with C-terminal-specific gastrin/CCK antisera, except in the proximal intestine where CCK-specific IR cells were absent (El Salhy 1984).

Gastrin/CCK cells have also been reported present in several teleost species, and are especially abundant in the proximal intestine (Langer *et al.* 1979; Noaillac-Depeyre and Hollande 1981; Rombout and Taverne-Thiele 1982; Holmgren *et al.* 1982). The exact identity of the gastrin/CCK/caerulein-like peptide present in teleosts has not so far been established (see later).

Glucagon and Insulin

These two peptides are present in the pancreatic islets in higher vertebrates. In lower vertebrates, however, they mainly occur in the gastro-intestinal tract, but moved gradually into the pancreas during evolution (see Van Noorden and Falkmer 1980). In the cyclostomes *Lampetra fluviatilis* and *Myxine glutinosa* glucagon IR cells are found in the intestinal mucosa only (Van Noorden and Pearse 1974; Falkmer and Östberg 1977), whereas insulin is found in the pancreatic islets and is thereby the first islet hormone (Van Noorden and Falkmer 1980). In *Squalus acanthias* endocrine cells containing glucagon IR are abundant in the intestine, but are also found scattered in other parts of the gastro-intestinal canal. In addition to glucagon, glicentin (the extended form of glucagon) has been observed in the stomach and middle intestine (El Salhy 1984).

In teleosts, glucagon IR cells are found in the intestine and pancreas (Langer *et al.* 1979; Noaillac-Depeyre and Hollande 1981; Holmgren *et al.* 1982; Rombout and Taverne-Thiele 1982; Rombout and Reinecke 1984), and in *Barbus conchonius* explicitly in both locations (Rombout and Reinecke 1984).

Insulin IR is reported to be present in pancreatic tissue and in Brockman's body in some teleosts (Langer *et al.* 1979).

Neurotensin

Neurotensin is a tridecapeptide present in the brain and in gut nerves and endocrine cells in both higher and lower vertebrates (see Reinecke, Carraway, Falkmer, Feurle and Forsmann 1980b). The physiological effects of neurotensin in mammals are mixed and include metabolic, vasomotor and gut motility responses (see

Hammer and Leeman 1981).

Neurotensin cells are very rare in *Myxine glutinosa*, where they occur in the intestine only (Reinecke *et al.* 1980b). In *Squalus acanthias* a low number of neurotensin IR cells have been found throughout the gut (Reinecke *et al.* 1980b; Holmgren and Nilsson 1983a; El Salhy 1984).

In teleost species, neurotensin IR cells are found throughout the gastro-intestinal canal, with the highest density in the proximal part of the gut (Reinecke *et al.* 1980a,b; Rombout and Reinecke 1984).

Somatostatin

Somatostatin is the second islet hormone (after insulin) to appear in the pancreas during evolution (see Van Noorden and Falkmer 1980). Somatostatin is, however, also found in neurons and endocrine cells along the gut, and a wide range of effects, including inhibition of secretion of pancreatic and gastro-intestinal hormones are described in mammals (see Patel, Zingg, Fitz-Patrick and Strikant 1981).

In *Myxine glutinosa* a few islet cells are found containing somatostatin IR, and some appear in the bile duct, but the main body of somatostatin IR cells are found in the intestinal mucosa (Van Noorden and Falkmer 1980; Reinecke, Falkmer, Heinrich, Almasan and Forsmann 1984). In *Squalus acanthias* numerous somatostatin IR cells are located in the gastric mucosa and a distally decreasing number in the intestinal mucosa (Figure 7.1B; Holmgren and Nilsson 1983a; El Salhy 1984; Reinecke *et al.* 1984).

The distribution of somatostatin IR cells in teleosts seems to vary between species. Several authors report the presence of somatostatin IR in the gastric mucosa (Noaillac-Depeyre and Hollande 1981; Langer *et al.* 1979; Holmgren *et al.* 1982; Reinecke *et al.* 1984). In *Myoxocephalus* (*Cottus*) *scorpius*, somatostatin IR cells occur also in the intestine and pancreas (Reinecke *et al.* 1984). In *Cyprinus carpio*, a stomachless fish, no somatostatin IR cells could be detected, whereas other stomachless fish, *Barbus conchonius*, has somatostatin cells in the pancreas (Rombout and Reinecke 1984).

Substance P

Substance P is an undecapeptide member of an 'expanding' family of peptides, the tachykinins, which all have similar amino acid sequences. Substance P possesses both endocrine and neuro-transmitter functions in mammals, and is, for example, considered to be the transmitter in sensory pain neurons (see Powell and Skrabanek 1981).

No substance P IR cells have been reported in the cyclostomes. In *Squalus acanthias*, on the other hand, substance P containing cells have been found: sparse in the stomach but numerous in the intestine and rectum (Holmgren and Nilsson 1983a; El Salhy 1984). Substance P IR cells are present also in a number of teleost species. The IR cells seem to occur throughout the gut, in some species especially abundant in the middle part of the gut and increasing towards the anus (Langer *et al.* 1979; Holmgren *et al.* 1982, Rombout and Reinecke 1984).

Vasoactive Intestinal Polypeptide

Vasoactive intestinal polypeptide (VIP) is a 28 amino acid peptide, first described as a vasodilatory agent, but with a wide range of additional functions on muscular tonus, gut secretions and meta-bolism in mammals (see Said 1981). The presence of VIP in nerves is well established, but the presence of VIP in endocrine cells has been disputed, in fish as in mammals. However, VIP has been reported to occur in endocrine cells in some fish species: Reinecke, Schluter, Yanaihara and Forsmann (1981) found a few cells in the intestinal epithelium of *Myxine glutinosa*, as well as in the whole gut of *Squalus acanthias*. VIP IR cells have also been reported in the rectum of *Squalus* (El Salhy 1984).

In teleosts, VIP IR cells have been described in the intestine of three species: *Cottus scorpius*, *Raniceps raninus* and *Barbus conchonius* (Reinecke *et al.* 1981; Rombout and Reinecke 1984).

Nerves in the Gastrointestinal Tract of Fish

The autonomic nerves of the gut are regarded as a separate part of the autonomic nervous system — the enteric nervous system (Langley 1921). The arrangement of the enteric nervous system in fish follows the general pattern of that in other vertebrates, with only minor variations (Kirtisinghe 1940; Nicol 1952). A well-developed nerve plexus — the myenteric plexus — containing ganglion cells, is present between the longitudinal (outer) and circular (inner) muscle layers. The innervation of the thin longitudinal muscle layer is comparatively sparse, whereas the circular muscle layer is densely innervated. A submucous plexus, also carrying ganglia, innervates the submucosa, often with an increased density towards the circular muscle. Fibres are seen in the muscularis mucosae and in the mucosa itself.

Immunohistochemistry has shown the presence of a number of peptides in nerves of the fish gut (Table 7.1). There are reports of bombesin-, gastrin/CCK-, somatostatin-, substance P- and VIP-IR in enteric nerves of an elasmobranch, *Squalus acanthias* (Holmgren and Nilsson 1983a; El Salhy 1984; Holmgren 1985). A holostean fish, *Lepisosteus platyrhincus*, has bombesin-, neurotensin- and VIP-IR nerves in the myenteric plexus and muscle layers (Holmgren and Nilsson 1983b). Several teleost species have been studied, and the presence of enteric nerves showing bombesin-, enkephalin-, gastrin/CCK-, neurotensin-, PHI (Peptide Histidine-Isoleucine)-, somatostatin-, substance P- (Figure 7.1c) and VIP-IR is reported (Langer *et al.* 1979, Van Noorden and Falkmer 1980; Van Noorden and Patent 1980; Holmgren *et al.* 1982; Reinecke *et al.* 1984; Rombout and Reinecke 1984; Thorndyke, Holmgren, Nilsson and Falkmer 1984; Jensen and Holmgren 1985). Of these peptides, VIP is reported present in almost all of the species studied so far. Bombesin-, substance P- and VIP-IRs are present in extensive networks in the myenteric and submucous plexuses and the circular muscle layer, and fibres also reach the mucosa (Holmgren *et al.* 1982; Holmgren and Nilsson 1983a,b; Holmgren 1985; Jensen and Holmgren 1985). There is thus a good morphological background to the assumption that these or closely related peptides are involved in the control of the gut functions in fish.

Neuropeptide Control of Gut Motility in Fish

The fish gut, just as the gut of other vertebrates, has the ability to receive food, process it physically and chemically, absorb nutrients and dispose of the wastes. These functions are intimately controlled by the enteric nervous system and by substances released from endocrine and paracrine cells of the gut mucosa. The control systems affect motility (even complex functions like peristalsis), secretion, regional blood flow, and possibly also other functions which are less well elucidated.

Judging from histochemical studies in fish and in comparison with mammalian experiences, a number of regulatory peptides are involved in the control of gut motility in fish, although so far such studies are comparatively sparse. Available information on the effects of different peptides on fish gut motility is summarised in Table 7.2. Experiments on isolated strip preparations of the gut wall or on the vascularly perfused stomach or intestine have shown that, for example, bombesin, substance P, enkephalin and neurotensin have an excitatory effect on gut smooth muscle from fish (Figure 7.2, references in Table 7.2). The excitatory response may be an increase in baseline tonus, an increase in amplitude and/or frequency of the rhythmic activity, or combinations of these effects (Figures 7.2A-E).

Inhibitory effects on the smooth muscle activity have been obtained with VIP in *Squalus* rectum and in the swimbladder muscle wall and vasculature in *Gadus morhua*. Bombesin is inhibitory in the perfused intestine of *G. morhua*, albeit in high concentrations (for references see Table 7.2).

The mechanisms of action of bombesin, substance P and VIP have been more closely studied and will now be discussed in more detail.

Bombesin

Bombesin is excitatory on both longitudinal and circular muscle from the stomach of *Salmo gairdneri* (Holmgren 1983), on muscle preparations from the rectum of *Squalus acanthias* (Lundin, Holmgren and Nilsson 1984) and on stomach strips from *Myoxocephalus* (*Cottus*) *scorpius* and *Gadus morhua* (Thorndyke *et al.* 1984). Immunohistochemistry shows the presence of bombesin IR in all these species (Holmgren and Nilsson 1983a; and unpublished; Thorndyke *et al.* 1984), although the character of the IR in *Salmo gairdneri* differs from that in the other three species.

Acetylcholine, like bombesin, is excitatory on the fish stomach. It was found that a combination of the two substances produced a potentiated contractile response of the fish stomach, i.e. the effect of the combined drugs exceeds the sum of the effects produced by bombesin or acetylcholine alone.

The potentiated effect is unaffected by tetrodotoxin or exchange of acetylcholine for the more cholinesterase-resistant drug

Table 7.2: The Effect of Putative Regulatory Peptides on Motility of Fish Gut Preparations

Bombesin	excitation	*Salmo g.* stomach	Holmgren 1983
		Squalus a. rectum	Lundin *et al.* 1984
		Gadus m. stomach	Thorndyke *et al.* 1984
		Myoxocephalus stomach	Thorndyke *et al.* 1984
	inhibition	*Gadus m.* intestine	Jensen and Holmgren 1985
Met-enkephalin	excitation	*Salmo g.* stomach	Holmgren 1983
	dual	*Gadus m.* intestine	Jensen and Holmgren 1985
Neurotensin	excitation	*Salmo g.* stomach	Holmgren 1983
Somatostatin	excitation	*Salmo g.* stomach	Holmgren 1983
	inhibition	*Gadus m.* intestine	Jensen and Holmgren 1985
	dual	*Squalus a.* rectum	Lundin *et al.* 1984
Substance P	excitation	*Salmo g.* stomach	Holmgren 1983
		Squalus a. gut	Holmgren 1985
		Gadus m. intestine	Jensen and Holmgren 1985
VIP	inhibition	*Squalus a.* rectum	Lundin *et al.* 1984
	dual	*Salmo g.* stomach	Holmgren 1983

carbachol, indicating that the effect is neither via a second neuron nor due to cholinesterase inhibition. The potentiation as well as the response to acetylcholine alone is blocked by atropine, and it is suggested that the effect of bombesin is somewhere at the synaptic site for the cholinergic neuron on the smooth muscle cell (Thorndyke *et al.* 1984).

Substance P

A substance P-like peptide was the first brain-gut peptide demonstrated in fish (von Euler and Östlund 1956). The effect of substance P and related peptides has now been studied in more detail. Substance P is excitatory on gut smooth muscle wherever it has been studied (Holmgren 1983, 1985; Jensen and Holmgren 1985). In *Squalus acanthias* the effect of substance P seems to be exclusively an excitatory effect directly on receptors on the smooth muscle cells (Holmgren 1985), while in *Salmo gairdneri* an additional excitatory mechanism via release of 5-hydroxytryptamine (5-HT) seems to be present (Holmgren *et al.* 1985). In *Salmo* the effect of low concentrations of substance P can be

Figure 7.2: Different Types of Excitatory Responses of Fish Gut Smooth Muscle Evoked by Exogenous Application of Regulatory Peptides (Horizontal Bars). A: Bombesin (10^{-7}M and 10^{-6}M) produces a concentration-dependent increase in both amplitude and frequency of rhythmic contractions in a strip preparation of circular muscle from the pyloric stomach of *Salmo gairdneri*; vertical bar = 6mN. B: Neurotensin (10^{-11}M and 10^{-10}M) superfused over longitudinal muscle strip preparations of cardiac stomach from *Salmo gairdneri* causes mainly an increase in tonus; vertical bar = 6 mN. C: Met-enkephalin (10^{-6}M) superfused over a strip preparation of circular smooth muscle from the pyloric stomach of *Salmo gairdneri* induces strong rhythmic activity superimposed on a small increase in basic tonus; vertical bar = 12 mN. D: A vascularly perfused whole stomach from the rainbow trout responds to substance P (10^{-8}M) with an increase in tension, reflected as an expulsion of fluid from the stomach; vertical bar = 10 ml. E: Substance P (10^{-8}M) induces rhythmic activity in a preparation of intestinal wall smooth muscle from *Squalus acanthias*; vertical bar = 20 mN

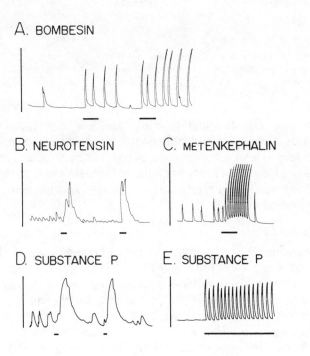

A. BOMBESIN

B. NEUROTENSIN

C. METENKEPHALIN

D. SUBSTANCE P

E. SUBSTANCE P

blocked by methysergide in concentrations that otherwise specifically block the effect of 5-HT both in isolated strip preparations and in the perfused whole stomach. Tetrodotoxin similarly blocks the effects of low concentrations of substance P, indicating that the effect is mediated by a second neuron, and the immunohistochemical findings of substance P mainly in the myenteric plexus (Holmgren *et al.* 1982) and of 5-HT in nerve structures further support the hypothesis. In addition it was shown that substance P releases label from a stomach preloaded with ³H-5-HT.

At least two types of substance P receptors have been described in mammals (Iversen, Hanley, Sandberg, Lee, Pinnock and Watson 1982): the SP-E receptor which is more sensitive to the tachykinin eledoisin, and the SP-P receptor which is approximately equally sensitive to the three tachykinins substance P, physalaemin and eledoisin. Experiments both in *Squalus* and in *Salmo* indicate that the substance P receptors mediating the effects on motility in the fish gut, if at all comparable to those present in mammals, more resemble the SP-P receptors.

VIP

VIP IR seems to be present in extensive nerve nets in the gut of all species studied. However, the physiological function of VIP has been difficult to establish. Experiments both in *Salmo gairdneri* and *Gadus morhua* have shown inconsistent responses, VIP sometimes inhibiting, sometimes exciting and often having no effect at all on the gut motility (Holmgren 1983; Jensen and Holmgren 1985). This may depend on structural differences between the naturally occurring fish VIP molecules and the porcine VIP available for the physiological experiments. It is also possible that the main function of VIP is as a cotransmitter or modulator, while the physiological experiments in fish, so far, have looked only for a straightforward direct effect on, for example, the muscle activity.

On the other hand, the motility of the *Squalus* gut is clearly inhibited by VIP in low concentrations (Lundin *et al.* 1984 and unpublished) and the flow through the swimbladder and the gas gland of the cod is slowly increased on exposure to VIP due to dilatation of the swimbladder and coeliac arteries (Lundin and Holmgren 1984).

Gastric Secretion in Fish

Of the great number of peptide hormones present in the mammalian gastro-intestinal tract, only a few (e.g. gastrin) are recognised as physiological regulators of gastric acid and pepsin secretion. Many are active upon exogenous administration, but their physiological role remains unclear, partly because of difficulties in investigating their paracrine or nervous release and actions. The peptides may exert their effects directly on the secretory cells and/or indirectly by modulating the release of other peptides or amines of endocrine or neurocrine origin (Konturek 1980). Whereas the literature dealing with the involvement of peptide hormones in mammalian gastric secretory processes is huge and rapidly increasing, such studies in fish have only begun. In this section, effects of exogenous peptides on gastric secretion in the cod, *Gadus morhua*, will be reviewed. Because heterologous peptides were used, mostly of mammalian and amphibian origin, and because the attained plasma levels were not measured, the presence or absence of secretory effects must not be taken as proof that the corresponding peptide is, or is not, a physiological regulator in the cod. Only the tentative labelling as a candidate regulator is warranted.

The experimental model is a pyloric-ligated, gastric-fistula fish with separate perfusion of the stomach and intestine. The intestinal perfusion (one part sea water, two parts distilled water) permits the fish to maintain a positive water balance and prevents the drinking occurring in intestinally unperfused, and consequently dehydrated fish. Under these conditions, basal acid secretion is stable for at least one week, and amounts to about 100 μmol $H^+kg^{-1}h^{-1}$. Basal acid secretion depends on an intact vagal supply, and is sensitive to atropine and histamine H_2-antagonists. Histamine and carbachol are agonists, stimulating the acid output to 400-600 μmol$H^+kg^{-1}h^{-1}$ (Holstein 1979; Holstein and Cederberg 1980, 1984).

In mammals, the principal acid-stimulating hormone is gastrin. The structurally related peptides, CCK and caerulein (of mammalian and amphibian origin, respectively) inhibit gastrin-stimulated acid secretion in man and dog, but are stimulatory in the basal state. These peptides are partial agonists (see Konturek 1980). Acid secretion in a dogfish is stimulated by porcine gastrin (see Vigna 1983), but gastrin as well as the octapeptide of CCK

(CCK 8) and caerulein are inhibitors of basal acid secretion in the cod (Holstein 1982). Caerulein is the most potent inhibitor ($ID_{50}=13\,pmol\,kg^{-1}h^{-1}$), and for all three peptides the sulphated species were more potent than the desulphated. Both gastrin and CCK are present in the gastro-intestinal tract of higher vertebrates, but their distribution in lower vertebrates is controversial. In the cod only one gastrin-related peptide was detected (Larsson and Rehfeld, 1977), while in the coho salmon gastrin-like activity was demonstrated in the stomach and CCK-like activity in the intestine (Vigna 1979). The antisecretory effects elicited by the exogenous peptides suggest that 'cod gastrin', which may be closely related to caerulein (Larsson and Rehfeld 1977), functions as a gastron, i.e. as a secretory antagonist, rather than being a gastrin in the classical sense. As pointed out by Vigna (1983), such a role would be a major departure from the tetrapod pattern of regulation, and alternative interpretations must be held open. For example, if the tested peptides are of low efficacy compared with the endogenous one, the results may be explainable in terms of partial agonism.

Bombesin, or a related peptide, is a candidate for the stimulant of cod acid secretion. Low doses of this frog-skin tetradecapeptide stimulate acid secretion in the cod (Holstein and Humphrey 1980) as well as in mammals. In the latter, stimulation depends on the release by bombesin of endogenous gastrin (Walsh *et al.* 1981), whereas in the cod a gastrin-independent mechanism is more likely. This assumption is based on the failure to detect any changes in plasma levels of gastrin IR during bombesin-stimulated acid secretion, and is in concert with the proposed role for 'cod gastrin' as an antagonist in the regulation of acid secretion. Bombesin significantly depressed plasma VIP IR in the cod, and porcine VIP is an inhibitor of basal acid secretion (Holstein and Humphrey 1980; Holstein 1983). It cannot be excluded, therefore, that at least part of the bombesin-evoked acid secretion results from the lowering of an endogenous, inhibitory tonus.

In the basal state the output of pepsin from the *in vivo* perfused cod stomach is $0.2\text{-}1.2\,mg\,kg^{-1}h^{-1}$, in terms of a standard pepsin (Merck, $100\,mU\,mg^{-1}$). If anything, the gastrin-related peptides inhibit pepsin output, and bombesin causes a slight stimulation of about 50 per cent (B. Holstein and C. Cederberg, unpublished observations). Histamine and carbachol, which are much more effective acid stimulators than bombesin, are weak pepsigogues, eliciting a pepsin output 100-200 per cent above basal (Holstein

Figure 7.3: Effects of Somatostatin on Histamine-induced Acid Secretion and Eledoisin-induced Pepsin Secretion in *Gadus morhua*, Measured in the Same Experiment. Histamine (H, 15 μg kg^{-1}h^{-1}) was i.m. infused alone (open columns) or together with somatostatin tetradecapeptide (S, 15 nmol kg^{-1}h^{-1}) (hatched columns) for 9h. The left-hand columns, which represent the incremental acid response over the first 4h, show a 77% inhibition by somatostatin of histamine-induced acid output. During the following 5h, eledoisin (E, 3.27 nmol kg^{-1}h^{-1}) was given to stimulate pepsin secretion. In the somatostatin-treated series, the 5h incremental pepsin output in response to eledoisin was depressed (not statistically significant) by 28%. Mean±SE of 10 control animals and 7 somatostatin-treated animals

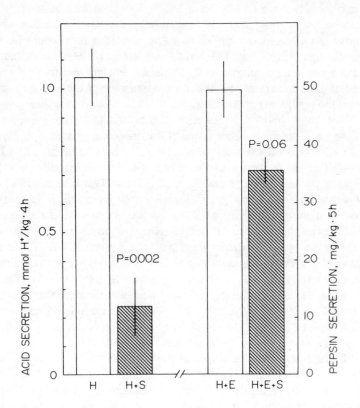

and Cederberg 1984). The response to histamine is transient and probably represents a wash-out phenomenon.

Peptides belonging to the tachykinin family are very active in stimulating pepsin secretion in the cod (Holstein and Cederberg 1985). Due to their high intrinsic activity (peak response about 10 mg pepsin $kg^{-1}h^{-1}$) and potency (ED_{50}=0.1 nmol $kg^{-1}h^{-1}$ for eledoisin and physalaemin, and 2.5 nmol $kg^{-1}h^{-1}$ for substance P), some tachykinin is probably involved in the regulation of pepsin secretion in the cod. Tachykinins stimulate enzyme secretion from mammalian pancreatic acini (Gardner and Jensen 1980); their effects on mammalian pepsin secretion appear not to have been investigated. High doses of physalaemin and eledoisin are inhibitory on cod acid secretion, whereas low doses tend to stimulate. Interestingly, a very similar response pattern is elicited by 5-HT; whereas high doses (>1 μmol $kg^{-1}h^{-1}$) inhibit and low doses stimulate acid secretion, pepsin secretion is stimulated over the entire dose range (Holstein and Cederberg 1984).

In mammals, somatostatin inhibits most endocrine and exocrine secretions from the gastro-intestinal tract and pancreas (for references, see Albinus *et al.* 1985). There is evidence that similar actions are also exerted in teleosts, e.g. insulin secretion in the eel (Ince 1980) and gastric acid secretion in the cod (Figure 7.3).

Obviously, secretions of acid and pepsin can easily be differentially affected in the cod (almost exclusive stimulation of acid by histamine; stimulation of pepsin/inhibition of acid by tachykinins and 5-HT). This can be achieved also in mammalian species where secretin inhibits (parietal cell) acid secretion and stimulates (chief cell) pepsin secretion (see Konturek 1980). However, in sub-mammalian vertebrates acid and pepsin are the products of a single (oxynticopeptic) cell population. Conceivably, separate sub-cellular pathways for the regulation of the two products are required.

References

Albinus, M., Gomez-Pan, A., Hirst, B.H. and Shaw, B. (1985) 'Evidence against Prostaglandin-mediation of Somatostatin Inhibition of Gastric Secretions', *Regulatory Peptides, 10,* 259-66

Anastasi, A., Erspamer, V. and Endean, R. (1968) 'Isolation and Structure of Caerulein, an Active Decapeptide from the Skin of *Hyla caerulea*', *Experientia, 23,* 699-700

Anastasi, A., Erspamer, V. and Bucci, M. (1971) 'Isolation and Structure of Bombesin and Alytesin, Two Analogous Active Peptides from the Skin of the European Amphibians *Bombina* and *Alytes*', *Experientia, 27,* 166-7

El Salhy, M. (1984) 'Immunocytochemical Investigation of the Gastro-entero-pancreatic (GEP) Neurohormonal Peptides in the Pancreas and Gastrointestinal Tract of the Dogfish, *Squalus acanthias*', *Histochemistry, 80,* 193-205

Euler, U.S. von. and Östlund, Y. (1956) 'Occurrence of a Substance P-like Polypeptide in Fish Intestine and Brain', *British Journal of Pharmacology, 11,* 323-5

Falkmer, S. and Östberg Y. (1977) 'Comparative Morphology of Pancreatic Islets in Animals', in B.W. Volk and K.F. Wellman (eds), *The Diabetic Pancreas,* Plenum, New York, pp. 15-59

Gardner, J.D. and Jensen, R.T. (1980) 'Receptor for Secretagogues on Pancreatic Acinar Cells', *American Journal of Physiology, 238,* G63-6

Hammer, R.A. and Leeman, S.E. (1981) 'Neurotensin: Properties and Actions', in S.R. Bloom and J.M. Polak (eds), *Gut Hormones,* Churchill Livingstone, Edinburgh, pp. 290-9

Holmgren, S. (1983) 'The Effect of Putative Non-adrenergic, Non-cholinergic Autonomic Transmitters on Isolated Strips from the Stomach of the Rainbow Trout, *Salmo gairdneri*', *Comparative and Biochemical Physiology, 74C,* 229-38

Holmgren, S. (1985) 'Substance P in the Gastro-intestinal Tract of *Squalus acanthias*', *Molecular Physiology, 8,* 119-30

Holmgren, S. and Nilsson, S. (1983a) 'Bombesin-, Gastrin/CCK-, 5-Hydroxytryptamine-, Neurotensin-, Somatostatin-, and VIP-like Immunoreactivity and Catecholamine Fluorescence in the Gut of the Elasmobranch, *Squalus acanthias*', *Cell and Tissue Research, 234,* 595-618

Holmgren, S. and Nilsson, S. (1983b) 'VIP-, Bombesin-, and Neurotensin-like Immunoreactivity in Neurons of the Gut of the Holostean Fish, *Lepisosteus platyrhincus*', *Acta Zoologica (Stockholm), 64,* 25-32

Holmgren, S., Vaillant, C. and Dimaline, R. (1982) 'VIP-, Substance P-, Gastrin/CCK-, Bombesin-, Somatostatin-, and Glucagon-like Immunoreactivity in the Gut of the Rainbow Trout, *Salmo gairdneri*', *Cell and Tissue Research, 223,* 141-53

Holmgren, S., Grove D.J. and Nilsson, S. (1985) 'Substance P Acts by Releasing 5-Hydroxytryptamine from Enteric Neurons in the Stomach of Rainbow Trout, *Salmo gairdneri*', *Neuroscience, 14,* 683-93

Holstein, B. (1979) 'Gastric Acid Secretion and Water Balance in the Marine Teleost, *Gadus morhua*', *Acta Physiologica Scandinavica, 105,* 93-107

Holstein, B. (1982) 'Inhibition of Gastric Acid Secretion in the Atlantic Cod, *Gadus morhua*, by Sulphated and Desulphated Gastrin, Caerulein, and CCK-octapeptide', *Acta Physiologica Scandinavica 114,* 453-9

Holstein, B. (1983) 'Effect of Vasoactive Intestinal Polypeptide on Gastric Acid Secretion and Mucosal Blood Flow in the Atlantic Cod, *Gadus morhua*', *General and Comparative Endocrinology, 52,* 471-3

Holstein, B. and Cederberg, C. (1980) 'Effect of Vagotomy and Glucose Administration on Gastric Acid Secretion in the Atlantic Cod, *Gadus morhua*', *Acta Physiologica Scandinavica*, *109*, 37-44

Holstein, B. and Cederberg, C. (1984) 'Effects of 5-HT on Basal and Stimulated Secretions of Acid and Pepsin and on Gastric Volume Outflow in the *in vivo* Gastrically and Intestinally Perfused Cod, *Gadus morhua*', *Agents and Actions*, *15*, 290-305

Holstein, B. and Cederberg, C. (1985) 'Effects of Substance P and Other Tachykinins on Gastric Acid and Pepsin Secretion, and on Gastric Volume Outflow in the Atlantic Cod, *Gadus morhua*', *American Journal of Physiology*, *250*, in press

Holstein, B. and Humphrey, C.S. (1980) 'Stimulation of Gastric Acid Secretion and Suppression of VIP-like Immunoreactivity by Bombesin in the Atlantic Codfish, *Gadus morhua*', *Acta Physiologica Scandinavica*, *109*, 217-23

Ince, B.W. (1980) 'Amino Acid Stimulation of Insulin Secretion from the *in situ* Perfused Eel Pancreas: Modification by Somatostatin, Adrenaline, and Theophylline', *General and Comparative Endocrinology*, *40*, 275-82

Iversen, L.L., Hanley, M.R., Sandberg, B.E.B., Lee, C-M., Pinnock, R.D. and Watson, S.P. (1982) 'Substance P Receptors in the Nervous System and Possible Receptor Subtypes', in R. Porter and M. O'Connor (eds), *Substance P in the Nervous System*, Ciba Foundation Symposium 91, Pitman, London, pp. 186-205

Jaffe, J.H. and Martin, W.R. (1975) 'Narcotic Analgesics and Antagonists', in L.S. Goodman and A. Gilman (eds), *The Pharmacological Basis of Therapeutics*, 5th edn, Macmillan, New York, pp. 245-54

Jensen, J. and Holmgren, S. (1985) 'Neurotransmitters in the Intestine of the Atlantic Cod, *Gadus morhua*', *Comparative and Biochemical Physiology*, *82C*, 81-9

Kirtisinghe, P. (1940) 'The Myenteric Nerve-plexus in Some Lower Chordates', *Quarterly Journal of Microscopical Sciences*, *81*, 521-39

Konturek, S.J. (1980) 'Gastrointestinal Hormones and Gastric Secretion', in G.B.J. Glass (ed.), *Gastrointestinal Hormones*, Raven Press, New York

Langer, M., Van Noorden, S., Polak, J.M. and Pearse A.G.E. (1979) 'Peptide Hormone-like Immunoreactivity in the Gastro-intestinal Tract and Endocrine Pancreas of Eleven Teleost Species', *Cell and Tissue Research*, *199*, 493-508

Langley, J.N. (1921) *The Autonomic Nervous System Part I*, Heffer, Cambridge

Larsson, L-I. and Rehfeld, J.F. (1977) 'Evidence for a Common Evolutionary Origin of Gastrin and Cholecystokinin', *Nature*, *269*, 335-8

Lundin, K. and Holmgren, S. (1984) 'Vasoactive Intestinal Polypeptide-like Immunoreactivity and Effects of VIP in the Swimbladder of the Cod, *Gadus morhua*', *Journal of Comparative Physiology*, *154B*, 627-33

Lundin, K., Holmgren, S. and Nilsson, S. (1984) 'Peptidergic Functions in the Dogfish Rectum', *Acta Physiologica Scandinavica*, *121*, 46A

McDonald, T.J., Nilsson, G., Vagne, M., Ghatei, M., Bloom, S.R. and Mutt, V. (1978) 'A Gastrin Releasing Peptide from the Porcine Non-antral Gastric Tissue', *Gut*, *19*, 767-74

Nicol, J.A.C. (1952) 'Autonomic Nervous Systems in Lower Chordates', *Biological Reviews*, *27*, 1-49

Noaillac-Depeyre, J. and Hollande, E. (1981) 'Evidence for Somatostatin-, Gastrin-, and Pancreatic Polypeptide-like Substances in the Mucosa Cells of the Gut of Fishes with and without Stomach', *Cell and Tissue Research*, *216*, 193-203

Östberg, Y., Van Noorden, S., Pearse, A.G.E. and Thomas, N.W. (1976) 'Cytochemical Immunofluorescence and Ultrastructural Investigation on

Polypeptide Hormone Containing Cells in the Intestinal Mucosa of a Cyclostome, *Myxine glutinosa'*, *General and Comparative Endocrinology, 28*, 213-27

Patel, Y.C., Zingg, H.H., Fitz-Patrick, D. and Srikant, C.B. (1981) 'Somatostatin: Some Aspects of its Physiology and Pathophysiology', in S.R. Bloom and J.M. Polak (eds), *Gut Hormones*, 2nd edn, Churchill Livingstone, Edinburgh, pp. 339-49

Powell, D. and Skrabanek, P. (1981) 'Substance P', in S.R. Bloom and J.M. Polak (eds), *Gut Hormones*, 2nd edn, Churchill Livingstone, Edinburgh, pp. 396-401

Reinecke, M., Almasan, K., Carraway, R., Helmstaedter, V. and Forsmann, W.G. (1980a) 'Distribution Patterns of Neurotensin-like Immunoreactive Cells in the Gastrointestinal Tract of Higher Vertebrates', *Cell and Tissue Research, 205*, 383-95

Reinecke, M., Carraway, R.F., Falkmer, S., Feurle, G. and Forsmann, W.G. (1980b) 'Occurrence of Neurotensin-immunoreactive Cells in the Digestive Tract of Lower Vertebrates and Deuterostomian Invertebrates', *Cell and Tissue Research, 212*, 173-83

Reinecke, M., Schluter, P., Yanaihara, N. and Forsmann, W.G. (1981) 'VIP Immunoreactivity in Enteric Nerves and Endocrine Cells of the Vertebrate Gut', *Peptides, 2*, 149-56

Reinecke, M., Falkmer, S., Heinrich, D., Almasan, K. and Forsmann, W.G. (1984) 'A Phylogenetic Study on the Occurrence of Somatostatin Immunoreactivity in Endocrine Cells and Enteric Nerves of the Vertebrate Gut', in S. Raptis, J. Rosenthal and J.E. Gerich (eds), *Proceedings of the 2nd International Symposium on Somatostatin*, 1-3 June 1985, Athens, Attempto Verlag, Tubigen pp. 47-53

Rombout, J.H.W.M. and Reinecke, M. (1984) 'Immunohistochemical Localization of (Neuro)peptide Hormones in Endocrine Cells and Nerves of the Gut of a Stomachless Teleost Fish, *Barbus conchonius* (Cyprinidae)', *Cell and Tissue Research, 237*, 57-65

Rombout, J.H.W.M. and Taverne-Thiele, J.J. (1982) 'An Immunocytochemical and Electron-microscopical Study of Endocrine Cells in the Gut and Pancreas of a Stomachless Teleost Fish, *Barbus conchonius* (Cyprinidae)', *Cell and Tissue Research, 227*, 577-93

Said, S.I. (1981) 'VIP Overview', in S.R. Bloom and J.M. Polak (eds), *Gut Hormones*, 2nd edn, Churchill Livingstone, Edinburgh, pp. 379-84

Solcia, E., Capella, C., Buffa, R., Frigerio, B., Usellini, L., and Fiocca, R. (1980) 'Morphological and Functional Classification of Endocrine Cells and Related Growth in the Gastrointestinal Tract', in G.B. Jerzy Glass (ed.), *Gastrointestinal Hormones*, Raven Press, New York, pp. 1-17

Sundler, F., Sjölund, K. and Håkansson, R. (1983) 'Gut Endocrine Cells — an Overview', *Uppsala Journal of Medical Sciences*, Suppl. 39

Thorndyke, M., Holmgren, S., Nilsson, S. and Falkmer, S. (1984) 'Bombesin Potentiation of the Acetylcholine Response in Isolated Strips of Fish Stomach', *Regulatory Peptides, 9*, 350

Van Noorden, S. and Falkmer, S. (1980) 'Gut Islet Endocrinology — some Evolutionary Aspects', *Investigative Cell Pathology, 3*, 21-35

Van Noorden, S. and Patent, G.J. (1980) 'Vasoactive Intestinal Polypeptide-like Immunoreactivity in Nerves of the Pancreatic Islet of Teleost Fish, *Gillichthys mirabilis*', *Cell and Tissue Research, 212*, 139-46

Van Noorden, S. and Pearse, A.G.E. (1974) 'Immunoreactive Polypeptide Hormones in the Pancreas and Gut of the Lamprey', *General and Comparative Endocrinology, 23*, 311-24

Vigna, S.R. (1979) 'Distinction between Cholecystokinin-like and Gastrin-like

Biological Activities Extracted from Gastrointestinal Tissues of Some Lower Vertebrates', *General and Comparative Endocrinology, 39,* 512-20

Vigna, S.R. (1983) 'Evolution of Endocrine Regulation of Gastrointestinal Function in Lower Vertebrates', *American Zoologist, 23,* 729-38

Walsh, J.H., Maxwell, V., Ferrari, J. and Varner, A.A. (1981) 'Bombesin Stimulates Human Gastric Function by Gastrin-dependent and Independent Mechanisms', *Peptides, 2,* 193-8

8 GASTRO-INTESTINAL PHYSIOLOGY: RATES OF FOOD PROCESSING IN FISH

David J. Grove

During the present decade, the widening scope of research on fish feeding and growth has led to the appearance of a stimulating range of review articles. Different authors have examined the functional anatomy of the gut, including nervous and hormonal control of muscle and glands, factors controlling appetite and food selection, nutrient requirements, the rates and efficiency of digestion and the consequent limitations that apply on the rates of growth, reproduction and metabolism that can be expected both in fish farms and in natural populations of fish (Rankin, Pitcher and Duggan 1983; Tytler and Calow 1985). The rapid accumulation of detailed, quantitative observations on a variety of fish species using a wide range of techniques inevitably changes and enriches the ideas of ichthyologists. This article is designed to show how rapidly ideas have changed in estimating the process of gastro-intestinal emptying in fish. The topic is important, since digestion rates have been included in models describing production in fishable populations (e.g. Andersen 1982) and cultured fish (e.g. From and Rasmussen 1984).

Reception of Food by the Alimentary Tract

The first reaction that occurs after a fish has ingested a meal is that the stomach or foregut distends, although the extent of this distension depends on the size of the prey item and the rate of ingestion of prey. In *Salmo gairdneri*, such distension leads to rhythmic contractions of the smooth muscle as enteric nerves are reflexly activated (Figure 8.1). The motor nerves appear to be of several types since neither atropine nor methysergide alone are sufficient to prevent the response. The reflex can be prevented by tetrodotoxin, or by perfusion with a combination of atropine *and* methysergide, suggesting that some nerves releasing acetylcholine and others releasing 5-hydroxytryptamine form the final motorneurone path-

Figure 8.1: *Salmo gairdneri.* When a reservoir attached to an intragastric balloon is raised, distension of the stomach is monitored by efflux of water. A: *In vivo*: an immediate distension to V_{res} is followed by a slow further relaxation of the stomach towards V_{max}. B: *In vitro*: reflex contractions induced by distension are blocked by intra-arterial perfusion of tetrodotoxin (TTX, 1 μM for 1 h)

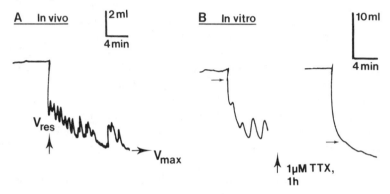

ways in the gut wall. Holmgren, Grove and Nilsson (1985) have shown that serotonergic nerves are present in the stomach and that substance P (presumably from interneurones) stimulates their activity. However, the gastric response to distension is more complex than this. The volume of the stomach begins to increase slowly after distension and this decrease in muscle tone remains for considerable periods after the stimulus is removed. Treatment with the polypeptides somatostatin and vasoactive intestinal polypeptide by themselves induce the relaxation and, since molecules of these types have been detected in gastric endocrine cells and in the gastric enteric plexus respectively, (Holmgren, Vaillant and Dimaline 1982), it is suggested that this receptive relaxation (to accommodate the volume of a meal) is influenced by gut hormones. Jobling (1982) reported that young *Pleuronectes* eat larger meals if food supply is restricted; Godin (1981) found a similar effect in *Oncorhynchus*. Part of this compensatory adaptation to restricted food supply may also be attributed to gastric relaxation, thereby allowing ingestion of larger meals. In the longer term, fish on low nutrient diets may increase the size of their stomach (Bromley and Adkins 1984). Once distension has taken place after a meal, the stomach or foregut initiates the process of disrupting and expelling the meal into the rest of the alimentary tract.

Gastric Emptying

Laboratory Studies

In 1979, Fänge and Grove collated a wide range of published data into their Figure 4 which clearly showed that the time required for a fish to empty its stomach depended on (a) temperature, and (b) the type of food taken by the fish in nature. Planktivorous fish empty their stomach in a shorter time than fish which eat larger items. Careful examination of the published data shows that gastric emptying time (GET) was already known to be effected by meal size and the size of the fish (see, e.g. Beamish 1972; Jobling, Gwyther and Grove 1977). Many workers had made more detailed studies of the rate at which the stomach empties by constructing emptying curves, in which the gastric emptying rate (GER) at any time could be measured (for methods see Talbot, 1985). Holmgren, Grove and Fletcher (1983) suggested that the shape of the emptying curve may well be predictable if a model is constructed incorporating the known biological and physiological variables. They suggested that an emptying curve for a particular food type would consist of two major phases:

(a) a temperature-sensitive delay time (t_d) during which newly ingested food would be ingested; and
(b) an emptying phase of duration t_{end} which depends on:

   (1) temperature,
   (2) the distension of the sac-like stomach,
   (3) the secretory surface area of the stomach, and
   (4) the surface area of the meal.

To formulate these ideas algebraically for predictive purposes, several assumptions were made. The effect of temperature on rate, within the normal physiological range, usually follows an exponential curve so that completion times decrease (e.g. $t_d = t'_d\, e^{-bT}$ where T is in °C and b is the exponent for the species in question). Stomach volume should be proportional to body weight, but the stretch measured by mechanoreceptors after ingestion of a meal is most likely to be a change in *linear* dimension in the wall of the expanding, three-dimensional sac. As a result of this *stimulus*, the stomach should secrete digestive juices in an amount proportional to the gastric surface area (the *response*). Such juices then act on the available surface area of the ingested food. Surprisingly, combination of these terms in a differential equation led to an *exponential* equation in which the amount of food (V_t) remaining in the stomach t hours into the emptying phase after feeding V_0 grams to a fish of weight W grams at T°C would be

$$\ln V_t = \ln V_0 - (K'e^{bT}W^{0.33})t$$

This conclusion does not imply that a given fish adjusts its gastric emptying rate simply by sensing the instantaneous bulk (V_t) in the stomach; the combination of stimulus, response and food surface area produces the equivalent curve. Many workers have found that this exponential curve gives an excellent fit to their experimental observations (see, for example, From and Rasmussen 1984). Finally, Holmgren *et al.* (1983) suggested that subdividing the initial meal (V_0) into n parts of identical shape, which remain separate, should decrease GET by increasing GER in a predictable way (GET proportional to $n^{-0.33}$).

Some of these predictions have been tested on the turbot (*Scophthalmus maximus*) (Grove, Moctezuma, Flett, Foott, Watson and Flowerdew 1985). Diets were prepared as firm, dry pellets which remained separate for much of the gastric digestion phase. In contrast with soft pellets (which rapidly fused *in vivo* to form a bolus), the discrete pellets were broken down at a rate governed by their surface area; increased subdivision of a standard food item increased GER. In addition, increase in meal size (achieved by feeding each fish with a variable number of standard pellets) led to a 'saturation' effect; presumably the gastric epithelium has a finite rate of secretion for which the pellets must compete. These studies allowed constants to be entered into a digestion equation which readily predicted both GER and GET, even for multiple items in the meal, and which also effectively predicted feeding rates in captive fish.

There remained, however, a problem which was at complete variance with the predictions of the above model. In *Scophthalmus, Limanda* and *Pleuronectes*, whether fed on natural food items or artificial diets of high quality, emptying curves were *not* found to be exponential, even when allowance was made for initial delays before food began to leave the stomach. Instead, the curves followed power equations close to those described by Jobling (1981b), notably his 'square-root' formulation in which

$$S_t^{0.5} = S_o^{0.5} - Kt$$

The logic involved in *predicting* that this relationship would be expected is weak, since it assumes that emptying rate only depends on the radial stretch of a very short segment of the stomach, rather than the three-dimensional stretching of a large sac. However, as with the exponential model described earlier, the final shape of the emptying curve must be attributable to an interaction of many factors which determine the secretory and mechanical activity rate of the stomach. A likely explanation which could lead to the well-substantiated 'square root' emptying curve was already documented in published research. Decrease in nutrient density of the fishes' food leads to increase in GER (e.g. Grove, Loizides and Nott 1978; Flowerdew and Grove 1979; Jobling 1981a). This response was attributed to nervous or hormonal negative-feedback mechanisms exerted on the stomach from the intestine when the latter contains nutrient-rich material. Such feedback is *not* included in any of the gastric emptying models published to date. Jobling (1986) points out that if such feedback causes the stomach to empty in *pulses,* then the shape of the emptying curve (estimated from discontinuous observations which form the majority of the techniques in use; see Talbot 1985) can vary according to the times at which observations are made. Even if the stomach empties continuously, inhibitory feedback can still modify the shape of the underlying exponential curve. The first chyme emptied from the stomach may be nutrient-rich but the later phases of emptying may release indigestible residua (see Karpevitch and Bokova 1936-7) which do not activate the inhibitory feedback.

In Figure 8.2 the non-exponential gastric emptying sequence in *Limanda* is shown using food pellets containing radio-opaque polystyrene/barium sulphate spheres (1 mm) and sufficient free barium sulphate powder mixed with the food to mark the different parts of the alimentary canal. Separate post-mortem studies on other individuals show that this species does not separate the individual spheres from the nutrient food as the stomach empties. In contrast, inclusion of radio-opaque iron particles (Figure 8.3) in the diet of young *Scophthalmus* leads to a selective retention of the particles relative to the food, seen in the X-radiograph as an increasing concentration of marker near to the pyloric sphincter. We have found that, in the wild, *Limanda* passes newly ingested, indigestible pieces of skeleton and shell through the pyloric

Figure 8.2: Gastric Emptying in *Limanda limanda*. The fish was fed with 2 g (1 % b.w.) of artificial flatfish diet containing 10 % (w/w) barium sulphate (to indicate shape of the gastro-intestinal tract) and barium sulphate impregnated polystyrene spheroids (*c.* 5 per food pellet) as an inert marker. X-rays were taken at 1, 5, 10 and 13 h after feeding at 18°C. In this fish, $S_t^{0.8} = S_o^{0.8} - 0.11\ t$, where S is in grams and t is in hours (S.C. Hailstone; unpublished data)

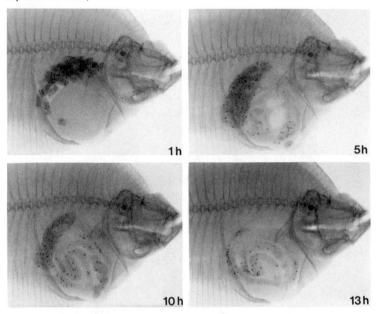

sphincter (see the first part of Figure 8.2 as well) whereas it appears that *Scophthalmus* does have a selective emptying sequence. Little is known about the factors that control the opening of the pyloric sphincter in fish. A major new observation (Figure 8.4) is that the flatfish pyloric sphincter possesses an adrenergic innervation originating in the sympathetic nervous system; catecholamines close the sphincter by activating alpha-adrenoceptors (J.H. Whillis, personal communication). There is as yet no information to show how this nerve supply is activated to control passage of food from the stomach, nor whether other nerves and/or hormones from the gut wall take part in the control of gastric emptying.

Clearly new experiments are required to study the emptying of the stomach using *continuous* methods of observation of the influx of material to the intestine. Even if the details of control are not

Figure 8.3: Selective Gastric Emptying in *Scophthalmus maximus*. Juvenile fish (*c.*18g) were fed a diet containing 10% BaSO$_4$ and iron particles (400 μm). Soon after feeding at 18°C, the iron particles have been separated from the rest of the food. These will not leave the stomach until the rest of the meal has been emptied (Morrison; unpublished observations)

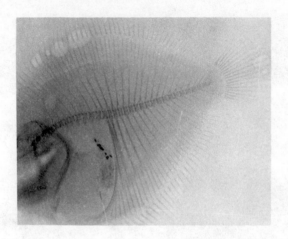

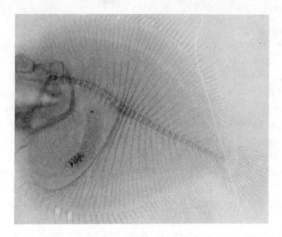

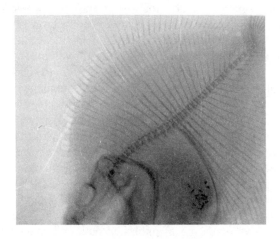

Figure 8.4: *Pleuronectes platessa*. Circular muscle of the pyloric sphincter prepared using the Falck-Hillarp histochemical technique for amines. Blue-green fluorescence in terminal varicose nerve fibres indicates the presence of catecholamines (J.H. Whillis; unpublished observations)

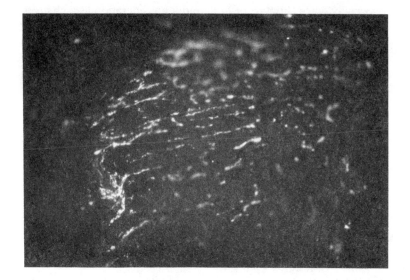

known, mathematical models for gastric emptying would need to include appropriate terms for a negative feedback on GER. Does the stomach release decreasing volumes of identical nutrient concentration as emptying proceeds? Or constant volumes of decreasing nutrient concentration? Or even does the stomach release varying volumes and concentrations in a time sequence such that (when a pulse is released) the intestine receives a constant mass of nutrient, dry weight or energy? Beamish (1972) has shown that different nutrient components of the meal appear in the intestine at different times. It is not impossible to foresee the formulation of suitable but complex algebraic expressions which include the 'feedback' component to describe the emptying process. At present, researchers are constrained to depend on empirical descriptions of their observations. Jobling (personal communication) has carried out an extensive reanalysis of published data and concluded that, in stated species, the *shape* of the emptying curve may not be constant. Changes from a basic 'exponential' curve through to power curves, or even linear emptying, may accompany changes in the physical and biochemical characteristics of the food (see also Brodeur, 1984). Extrapolation of digestion rates observed in laboratory studies to field conditions could be erroneous, especially when the type of food under consideration also changes. It is fortuitous, although gratifying, that the emptying rates of natural food items containing shells and cuticles (single meals) in *Limanda* (Fletcher, Grove, Basimi and Ghaddaf, 1984) and larger *Pleuronectes* (Basimi and Grove 1985) in the laboratory were closely predicted by empirical predictions based on artificial diets of high-nutrient density.

Much of the published work quantifying emptying rate in fish is, however, based on members of a stated species which have become structurally mature. Hofer and Uddin (1985) and Buddington (1985) emphasise that the young of such species (especially larvae) may lack organ systems and enzymes appropriate for the adult diet; early stanzas in the life history of a stated species require their own descriptions of clearance of food through the alimentary tract.

From the above, it may be expected that, despite incomplete understanding of the basic physiology of fish gastro-intestinal functioning, reasonable descriptions and predictions of gastric emptying can be obtained by recording the temperature, fish size, food type, meal size, energy content and particle size for the species under study. However, even these factors are not enough. Goddard (1974) examined the effects of long-term food deprivation in *Pleuronectes* and showed that gastric emptying rate decreased when the fish were offered a single meal. The previous feeding history of a fish is important if digestion rates are to be predicted in, say, wild fish populations after a period of fasting during winter and/or after a spawning season. Talbot (1985) reviewed his more detailed studies in which GER varies dramatically between fish fed a single, isolated meal and those fed multiple meals. Fletcher *et al.* (1984) found that multiple feeding in *Limanda* led to a smaller, but still significant, increase in gastric emptying rate. They also found that, if food items remain separate, the earliest ingested item is rapidly and preferentially emptied from the stomach. These observations are of particular importance when attempts are made to estimate fish digestion rate, and hence the rate of food intake, in populations outside the laboratory.

Field Studies

Soofiani and Hawkins (1985) maintain that it is easy to detect the nature and quality of food consumed by fish in the wild, but that the rate of intake is more difficult to assess. They review the problems encountered in obtaining suitable samples and the likely sources of error which are encountered when estimating food intake by indirect means. Bearing their recommendations in mind, the conversion of gut contents in samples taken from natural populations into consumption rates depends on estimates of the rate of gastro-intestinal processing. Observed individual prey items in the stomach or foregut can be given a code, representing the visual appearance of the item depending on the state of digestion (e.g. Hofsten, Kahan, Katznelson and Bar-El 1983). Laboratory studies can determine the *time* required for such digestion stages to be reached, bearing in mind the temperature, fullness and predator size. Knowledge of the mass or energy content of each prey item from field samples allows food intake to be estimated per unit time (Swenson and Smith 1973). In a second method, where food items are not easily distinguished, sequential samples from the fish population are taken and stomach contents assessed (e.g. as dry weights). Basimi and Grove (1985) found from laboratory studies that the emptying curve for *Pleuronectes* could be described by the formula:

$$S_t^b = S_0^b - Kt$$

where $b = 0.75$ and $K = 0.0068 W^{0.43} e^{0.041\ T} (W =$ fish weight and T is water temperature in °C). Field studies of stomach contents in different seasons for different weights of fish in the population at known temperature, taken every 3 h during the day by trawl, allowed two estimates of food intake F_t to be made:

(a) assuming no digestion occurred for any item ingested in the period between samples,
$F_t = S_2 - (S_1^b - Kt)^{1/b}$ where S_1 and S_2 are the observed contents in samples at the beginning and end of the 3-hour period (minimum estimate);
(b) assuming that all newly ingested items were consumed exactly halfway through the period ($t/2 = 1.5$ h) and that gastric emptying of such items began immediately,
$F_t = (S_2^b + (Kt)/2)^{1/b} - (S_1^b - (Kt)/2)^{1/b}$
(maximum estimate).

In fact, since gastric emptying time (even in summer) is large relative to the intersampling interval used here, the two methods gave similar results. Studies of abundance enabled the annual consumption of the major food items (*Pectinaria* and *Abra*) by the fish population to be assessed. The population of half a million recruited fish consumed 2 g of prey per square metre during the growth season from March to July.

Hofer, Forstner and Rettenwander (1982) developed a further technique, appropriate for fish in lakes or similar restricted water habitats. Textile dyes were used to label cellulose powder in an attractive artificial food which was thrown to wild roach. The fish were captured at a known time after feeding, using rods or nets, and the feeding rate was taken as the dry weight of food which lay anterior to the red marker in the intestine of this stomachless fish. Such field experiments produced results which agreed closely with laboratory observations of fish under *ad libitum* feeding.

The problems that face fish researchers in this subject are of widespread importance. Analysis of food supply and resource partitioning among competing species in a fishery is as yet rudimentary. The effects of predation by one commercial species on another influence estimates of natural mortality in fishery assessment models. The consequences of changes in food supply caused by natural and man-made impacts on the environment cannot yet be predicted with any confidence, save in restricted bodies of water of relatively simple biological structure. Field studies of production of fish require, among many other inputs, improved predictions of the rate at which fish process food in natural populations. Such predictions have improved in recent years, as laboratory studies have increased in quality. However, our ability to model fish feeding in nature still falls far short of the precision necessary for predicting how the component species of an ecosystem will fare as a result of physical and biological fluctuations, including over exploitation of commercial species by directed fishing activities.

References

Andersen, K.P. (1982) 'An Interpretation of the Stomach Contents of a Fish in Relation to Prey Abundance', *Dana*, 2, 1-50

Basimi, R.A. and Grove, D.J. (1985) 'Gastric Emptying Rate in *Pleuronectes platessa*', *Journal of Fish Biology*, 26, 545-52

Beamish, F.H.W. (1972) 'Ration Size and Digestion in Large-mouth Bass, *Micropterus salmonides* L.', *Canadian Journal of Zoology*, 50, 153-64

Brodeur, R.D. (1984) 'Gastric Evacuation Rates for Two Foods in the Black Rockfish, *Sebastes melanops* Girard', *Journal of Fish Biology*, 24, 287-98

Bromley, P.J. and Adkins, T.C. (1984) 'The Influence of Cellulose Filler on Feeding, Growth and Utilization of Protein and Energy in Rainbow Trout, *Salmo gairdneri* Richardson', *Journal of Fish Biology*, 24, 235-44

Buddington, R.K. (1985) 'Digestive Secretions of Lake Sturgeon, *Acipenser fulvescens*, during Early Development', *Journal of Fish Biology*, 26, 715-23

Fänge, R. and Grove, D.J. (1979) 'Digestion', in W.S. Hoar, D.J. Randall and J.R. Brett (eds), *Fish Physiology* vol. VIII, Academic Press, New York, pp. 161-260

Fletcher, D.J., Grove, D.J., Basimi, R.A. and Ghaddaf, A. (1984) 'Emptying Rates of Single and Double Meals of Different Food Quality from the Stomach of the Dab, *Limanda limanda* (L.)', *Journal of Fish Biology*, 25, 435-44

Flowerdew, M.W. and Grove, D.J. (1979) 'Some Observations on the Effects of Body Weight, Temperature, Meal Size and Quality on the Gastric Emptying Time in Turbot, *Scophthalmus maximus* (L.) using Radiography', *Journal of Fish Biology*, 14, 229-38

From, J. and Rasmussen, G. (1984) 'A Growth Model, Gastric Evacuation, and Body Composition in Rainbow Trout, *Salmo gairdneri* Richardson, 1836', *Dana*, 3, 61-139

Goddard, J.S. (1974) 'An X-ray Investigation of the Effects of Starvation and Drugs on Intestinal Motility in the Plaice, *Pleuronectes platessa* (L.)', *Ichthyologica*, 6, 49-58

Godin, J.-G.J. (1981) 'Effect of Hunger on the Daily Pattern of Feeding Rates in Juvenile Pink Salmon, *Oncorhynchus gorbuscha* Walbaum', *Journal of Fish Biology*, 19, 63-71

Grove, D.J., Loizides, L. and Nott. J. (1978) 'Satiation Amount, Frequency of Feeding and Gastric Emptying Rate in *Salmo gairdneri*', *Journal of Fish Biology*, 12, 507-16

Grove, D.J., Moctezuma, M.A.-H., Flett, R.J., Foott, J.S., Watson, T. and
Flowerdew, M. (1985) 'Gastric Emptying and the Return of Appetite in
Juvenile Turbot, *Scophthalmus maximus* (L.), Fed on Artificial Diets', *Journal
of Fish Biology, 26,* 339-54

Hofer, R. and Uddin, N.A. (1985) 'Digestive Processes during the Development of
the Roach, *Rutilus rutilus* (L.)', *Journal of Fish Biology, 26,* 419-28

Hofer, R., Forstner, H. and Rettenwander, R. (1982) 'Duration of Gut Passage
and its Dependence on Temperature and Food Consumption in Roach *Rutilus
rutilus* (L.): Laboratory and Field Experiments', *Journal of Fish Biology, 20,*
289-99

Hofsten, A.V., Kahan, D., Katznelson, R. and Bar-El, T. (1983) 'Digestion of
Free-living Nematodes Fed to Fish', *Journal of Fish Biology, 23,* 419-28

Holmgren, S., Vaillant, C. and Dimaline, R. (1982) 'VIP-, Substance P-,
Gastrin/CCK-, Bombesin-, Somatostatin- and Glucagon-like Immunoreactivity
in the Gut of the Rainbow Trout, *Salmo gairdneri', Cell and Tissue Research,
223,* 141-53

Holmgren, S., Grove, D.J. and Fletcher, D.J. (1983) 'Digestion and the Control of
Gastrointestinal Motility', in J.C. Rankin, T.J. Pitcher, and R. Duggan (eds),
Control Processes in Fish Physiology, Croom Helm, London, pp. 23-40

Holmgren, S., Grove, D.J. and Nilsson, S. (1985) 'Substance P Acts by Releasing
5-Hydroxytryptamine from Enteric Neurons in the Stomach of the Rainbow
Trout, *Salmo gairdneri', Neuroscience, 14,* 683-93

Jobling, M. (1981a) 'Dietary Digestibility and the Influence of Food Components
on Gastric Evacuation in Plaice, *Pleuronectes platessa* (L.)', *Journal of Fish
Biology, 19,* 29-36

Jobling, M. (1981b) 'Mathematical Models of Gastric Emptying and the
Estimation of Daily Rates of Food Consumption for Fish', *Journal of Fish
Biology, 19,* 245-57

Jobling, M. (1982) 'Some Observations on the Effects of Feeding Frequency on
the Food Intake and Growth of Plaice, *Pleuronectes platessa* (L.)', *Journal of
Fish Biology, 20,* 431-44

Jobling, M. (1986) 'Mythical Models of Gastric Emptying and Implications for
Food Consumption Studies', Gutshop 1984, *Environmental Biology of Fish,* in
press

Jobling, M., Gwyther, D. and Grove, D.J. (1977) 'Some Effects of Temperature,
Meal Size and Body Weight on Gastric Evacuation Time in the Dab, *Limanda
limanda* (L.)', *Journal of Fish Biology, 10,* 291-8

Karpevitch, A.F. and Bokova, E.N. (1936-7) 'The Rate of Digestion in Marine
Fish': Part 1, *Zoologisches Zeitschrift, 15,* 143-8; Part 2, *Zoologisches
Zeitschrift, 16,* 21-44

Rankin, J.C., Pitcher, T.J. and Duggan, R. (eds) (1983) *Control Processes in Fish
Physiology,* Croom Helm, London, 298 pp.

Soofiani, N.M. and Hawkins, A.D. (1985) 'Field Studies of Energy Budgets', in P.
Tytler and P. Calow (eds), *Fish Energetics: New Perspectives,* Croom Helm,
London and Johns Hopkins University Press, Baltimore, pp. 283-307

Swenson, W.A. and Smith, L.L. (1973) 'Gastric Digestion, Food Consumption,
Feeding Periodicity and Food Conversion Efficiency in Walleye, (*Stizostedion
vitreum vitreum*)', *Journal of the Fisheries Research Board of Canada, 30,*
1327-36

Talbot, C. (1985) 'Laboratory Methods in Fish Feeding and Nutritional Studies', in
P. Tytler and P. Calow (eds) *Fish Energetics: New Perspectives,* Croom Helm,
London, and Johns Hopkins University Press, Baltimore, pp. 125-54

Tytler, P. and Calow, P. (eds) (1985) *Fish Energetics: New Perspectives,* Croom
Helm, London, and Johns Hopkins University Press, Baltimore, 349 pp.

9 FILTRATION IN THE PERFUSED HAGFISH GLOMERULUS

Jay A. Riegel

There is now abundant experimental evidence that the mechanism of glomerular function in vertebrates in general is that proposed first by Starling: an excess of hydrostatic pressure causes outward filtration of fluid across the glomerular capillary wall, overcoming the fluid-attracting properties (colloid osmotic pressure: COP) of the blood proteins. Glomeruli of hagfish exhibit few apparent differences from glomeruli of other vertebrates either in their structure (e.g. Heath-Eves and McMillan 1974; Kühn, Stolte and Reale 1975) or overall function (Riegel 1978; Alt, Stolte, Eisenbach and Walvig 1981). Because of this close similarity, it would be expected that the functional mechanism of hagfish glomeruli and the glomeruli of the other vertebrates would be identical. However, as judged by conditions in the general blood circulation, pressure filtration is not possible in the hagfish. The COP of the blood plasma exceeds the hydrostatic pressure of the blood. Nevertheless, despite the apparently unfavourable pressure gradient, hagfish glomeruli form urine at an appreciable rate (Riegel 1978).

In more recent studies hydrostatic pressures in the hagfish have been re-examined using a servo-nulling pressure microtransducer (Riegel 1986). It was found that hydrostatic pressure in the glomerular capillaries attains a maximum value of $c.4\,cmH_2O$, although the average value measured was closer to $2\,cmH_2O$. The average value of the blood COP was found to be $14.5\,cmH_2O$, whereas the COP of glomerular fluid was negligible. These studies confirmed the conclusion formed earlier: hydrostatic pressure, derived from the arterial pulse, is not adequate to account for glomerular filtration in the hagfish.

Manipulations designed to alter the hydrostatic pressure in the renal vasculature can profoundly affect the filtration rate of perfused hagfish glomeruli (Stolte and Eisenbach 1973; Riegel 1978; Alt *et al.* 1981). If arterial pressure does not underlie glomerular filtration, why are hagfish glomeruli apparently so responsive to pressure? To answer this question recourse was made to the study of single glomeruli which were isolated from their blood supply and perfused with a colloid-containing Ringer. Simultaneously, perfusion pressure, perfusion rate, single glomerulus filtration rate (SGFR) and pressure in a glomerular capillary (P_{GC}) were monitored. The results of these experiments will be reported in detail elsewhere, but the demonstration of the independence of SGFR and P_{GC} is presented here.

Figure 9.1 summarises an experiment in which a renal corpuscle was isolated and perfused through its segmental artery using

Figure 9.1: Summary of the Results when an Isolated Hagfish Renal Corpuscle Was Perfused with Colloid Ringer through an Adjacent Segmental Artery. Measurements were made of the rate of perfusion, perfusion pressure, single glomerulus filtration rate (SGFR) and hydrostatic pressure in a glomerular capillary (P_{GC})

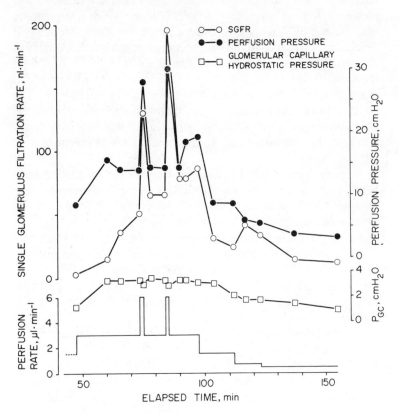

methods devised by Riegel (1978). Initially, the perfusion rate was 1.5 μl min^{-1} and the SGFR was very low. Measurements were begun of pressure in a capillary of the perfused glomerulus using a servo-nulling microtransducer (Riegel 1986), and the perfusion rate was increased to 3 μl min^{-1}. This action doubled the perfusion pressure to a value approximating those recorded in segmental arteries of anaesthetised hagfishes (Riegel 1986). At that time, pressure in the glomerular capillary rose to a value of 3.3 cmH$_2$O. For *c.*50 min, the perfusion rate was maintained at 3 μl min^{-1} except for two brief periods which will be discussed below. The SGFR rose steadily during the 50-min period and pressure in the glomerular capillary was maintained at about 3.3 cmH$_2$O. During periods of 1.5 min and 1 min, the perfusion rate was doubled to 6 μl min^{-1}. This resulted in a momentary doubling of the perfusion pressure and increases of the SGFR by about two to three times. However, hydrostatic pressure in the glomerular capillary actually fell by a small amount. It must be concluded from this observation that SGFR is independent of P_{GC} in the hagfish.

The experiment illustrated in Figure 9.1 is one of four which, although they differed in detail, all confirmed that hydrostatic pressure does not rise above 4 to 5 cmH$_2$O in glomerular capillaries even when high rates of perfusion prevailed. Furthermore, in 14 measurements of glomerular capillary hydrostatic pressure in anaesthetised hagfish, the highest value recorded was 4.2 cmH$_2$O. The SGFR of glomeruli whose blood supply was intact and which were undisturbed by pressure-measuring probes was comparable to the values shown in Figure 9.1 (Riegel 1986). Therefore, there is no reason to believe that Ringer-perfused glomeruli respond significantly differently to pressure changes in their perfusion system than do glomeruli of intact hagfish.

Glomerular filtration in the hagfish must depend upon the amount of fluid that is delivered to the glomerular capillaries. As demonstrated elsewhere (Riegel 1986), hagfish glomeruli are shunted by a low-resistance vascular pathway. Therefore, an unknown and probably variable proportion of the fluid entering the renal corpuscle does not flow through the glomerular capillaries. Consequently, when the perfusion rate was increased or decreased, SGFR increased or decreased because of changes in the volume of fluid in the capillaries, not changes in the capillary hydrostatic pressure. That is, glomerular filtration appears to be flow dependent.

In glomeruli of other animals the glomerular capillary pressure may fall to values where there is a pressure equilibrium across the capillary wall. When this occurs, glomerular filtration becomes flow dependent (Brenner, Baylis and Deen 1976). However, this condition is not seen as a permanent feature affecting the entire volume of the glomerulus. In some colonies of the so-called 'Munich-Wistar' strain of laboratory rats and in the squirrel monkey, filtration-pressure equilibrium may affect part of the capillary area (Arendshorst and Gottschalk 1985). The same is true of the amphibian, *Amphiuma means* (Persson 1981), and the lamprey, *Lampetra fluviatilis*, when it is acclimated to sea water isosmotic with its body fluids (McVicar and Rankin 1985).

The results summarised in Figure 9.1 provide only a partial answer to the question posed above: why is SGFR of hagfish so responsive to perfusion pressure? The SGFR rises in direct response to an increased flow through the glomerular capillaries. However, the connection between increased flow rate and SGFR in the absence of an effective filtration pressure remains a mystery. It must be concluded that some cellular mechanism is responsible for SGFR of hagfish glomeruli. As demonstrated by the writer (Riegel 1978), SGFR of perfused glomeruli was reversibly halted by the chemical inhibitors, ouabain and 2,4-dinitrophenol, although there was no detectable effect on perfusion pressure. This raised the possibility that active transport of some kind might be involved in transcapillary movement of fluid; this possibility still remains. However, the recent demonstration of the existence of a vascular pathway that shunts the glomerular capillaries weakens the argument in favour of active transport. It is possible that perfusion fluid was shunted away from the glomerular capillaries with no effect on pressure that was measurable in the perfusion line. Perhaps the inhibitors acted by affecting the musculature of the blood vessels supplying the glomerular capillaries, rather than a fluid-transporting epithelium. Clearly, much further work remains to be done before the enigma of the hagfish glomerulus can be solved.

Acknowledgements

It is a pleasure to dedicate this brief essay to Ragnar Fänge. Ragnar made it possible for me to commence study of pirålen by his kind invitations to visit Göteborg and his unstinting and ever helpful advice based on long years of acquaintanceship with the beast. This work was supported in part from the Central Research Fund of London University. The study was carried out at the Hopkins Marine Station of Stanford University, Pacific Grove, California, USA.

References

Alt, J.M., Stolte, H., Eisenbach, G.M. and Walvig, F. (1981) 'Renal Electrolyte and Fluid Excretion in the Atlantic Hagfish *Myxine glutinosa*', *Journal of Experimental Biology*, *91*, 323-30

Arendshorst, W.J. and Gottschalk, C.W. (1985) 'Glomerular Ultrafiltration Dynamics: Historical Perspective', *American Journal of Physiology*, *248*, F163-74

Brenner, B.M., Baylis, C. and Deen, W.M. (1976) 'Transport of Molecules across Renal Glomerular Capillaries', *Physiological Reviews*, *56*, 502-34

Heath-Eves, M.J. and McMillan, D.B. (1974) 'The Morphology of the Kidney of the Atlantic Hagfish, *Myxine glutinosa* (L.)', *American Journal of Anatomy*, *139*, 309-34

Kühn, K., Stolte, H. and Reale, E. (1975) 'The Fine Structure of the Kidney of the Hagfish (*Myxine glutinosa* L.)', *Cell and Tissue Research*, *164*, 201-13

McVicar, A.J. and Rankin, J.C. (1985) 'Dynamics of Glomerular Filtration in the River Lamprey, *Lampetra fluviatilis* L.', *American Journal of Physiology*, *249*, F132-8

Persson, E. (1981) 'Dynamics of Glomerular Ultrafiltration in *Amphiuma means*', *Pflügers Archiv*, *391*, 135-40

Riegel, J.A. (1978) 'Factors Affecting Glomerular Function in the Pacific Hagfish *Eptatretus stouti* (Lockington)', *Journal of Experimental Biology*, *73*, 261-77

Riegel, J.A. (1986) 'Hydrostatic Pressures in Glomeruli and Renal Vasculature of the Hagfish *Eptatretus stouti* (Lockington)', MS in preparation

Stolte, H. and Eisenbach, G.M. (1973) 'Single Nephron Filtration Rate in the Hagfish *Myxine glutinosa*', *Bulletin of the Mount Desert Island Biological Laboratory*, *13*, 120-1

10 PHYSIOLOGICAL METHODS IN FISH TOXICOLOGY: LABORATORY AND FIELD STUDIES

Lars Förlin, Carl Haux, Tommy Andersson, Per-Erik Olsson and Åke Larsson

During the last two decades, the use of biochemical and physiological methods in fish toxicology has expanded dramatically. This review provides some background on the use of a physiological approach to detect early and sublethal responses of fish to toxic substances. We have also described results of field and laboratory studies, and have discussed the advantages and problems associated with various techniques based on our experiments.

Sublethal Physiological Responses

The effects of pollutants can be studied at various levels of biological organisation, from the subcellular and individual level to alterations at the population and community levels (Table 10.1). Primarily, sublethal responses start as biochemical effects at the

Table 10.1: Effects of Toxicants on Different Levels of Biological Organisation

	Toxicant effects
Subcellular and cellular	Interaction with nucleic acids, enzymes, membrane structures or other functional components in the cell. These primary effects induce sequences of structural and functional alterations
Tissue and organ	Effects on vital functions, associated with nerve and muscle functions, respiration, circulation, immune defence, osmoregulation, and hormonal regulation
Individual	Effects on integrated functions which include behaviour, growth, reproduction and survival. Effects on population and biological community levels

158

subcellular or cellular level which, in turn, may induce a sequence of structural and functional alterations at a higher level of organisation. Such changes are often manifested by the impairment of vital physiological functions, e.g. nerve and muscle functions, respiration, circulation, immune defence, osmoregulation, and hormonal regulation (Larsson, Haux and Sjöbeck 1985). Ultimately, such effects can lead to irreversible and detrimental disturbances of integrated functions. Thus, the effects of pollutants on a population can be better understood and predicted by studying the sublethal effects on the individual, or by focusing on processes at lower levels of biological organisation. It is therefore possible to use sublethal physiological responses to detect pollutant-caused disturbances at a very early stage, since individual responses always precede population responses. Biochemical and physiological methods of diagnosis constitute a promising approach to the problem of detecting the effects of toxic chemicals at the earliest possible stage (Larsson *et al.* 1985).

Laboratory studies have shown many of these methods to be useful in diagnosing anomalies at low exposure levels. In particular, chronic exposure of fish to low levels of cadmium (Larsson 1975; Johansson-Sjöbeck and Larsson 1978; Larsson, Bengtsson and Haux 1981; Larsson and Haux 1982), lead (Johansson-Sjöbeck and Larsson 1979; Haux and Larsson 1982), chlorinated hydrocarbons (Förlin, Hansson, Haux, Johansson-Sjöbeck, Larsson and Lidman 1979; Haux, Larsson, Lidman, Förlin, Hansson and Johansson-Sjöbeck 1982), municipal effluents (Förlin and Hansson 1982a) and industrial effluents (Lehtinen, Larsson and Klingstedt 1984; Förlin, Andersson, Bengtsson, Härdig and Larsson, 1985) has revealed sublethal effects on various response parameters, including haematology, ion regulation, carbohydrate metabolism and the mixed function oxidase system.

The application of biochemical and physiological methods in field experiments on fish from polluted waters has also revealed pronounced effects of pollutants on several response parameters (Larsson, Haux and Sjöbeck 1984; Larsson *et al.* 1985; Förlin *et al.* 1985; Haux, Larsson, Lithner and Sjöbeck 1986). Sublethal effects obtained in field studies must, however, be viewed with caution as several of the biochemical and physiological response parameters are influenced by abiotic and biotic factors, e.g. sex, stage of sexual development, nutritional state, temperature, season. In addition, stress effects can be induced by the capture, holding and sampling procedures. Some of our results and experiences based on the application of biochemical and physiological methods in laboratory and field investigations are presented below. We have made a special effort to gauge the usefulness of each response parameter under both laboratory and field conditions.

The Same Response Parameters in the Laboratory and Field

Introduction

In our toxicological investigations on fish, the study of the same sublethal response parameters in the laboratory and field is primarily meant to help to overcome the difficulties encountered when interpreting the results of field experiments. The laboratory studies have provided the background upon which we base our understanding and interpretation of the responses measured in fish living in polluted rivers, lakes and coastal areas.

Methodological Aspects

We have devoted considerable effort to the development of the experimental techniques used in our investigations. These procedures, all of which include capturing of the fish with gill nets, careful handling, 2 to 4 days of recovery in fish chests, and sampling of tissues, have been described elsewhere (Dave, Johansson-Sjöbeck, Larsson, Lewander and Lidman 1975; Förlin and Andersson 1985; Haux, Larsson and Sjöbeck 1985), and will not be repeated here.

In our laboratory studies, environmental factors such as temperature, food supply and photoperiod have been shown to influence many biochemical and physiological parameters (Andersson, 1985; Larsson *et al.* 1985); Björnsson, Haux, Förlin and Deftos 1986). We have also shown that biological factors such as sex, stage of sexual development and hormone treatments can influence some of the response parameters (Förlin and Hansson 1982b; Hansson, Förlin, Rafter and Gustafsson 1982; Larsson *et al.* 1985; Björnsson *et al.* 1986). In feral fish populations, the combined influences of external and internal factors are known to cause background and seasonal variations (for references see Larsson *et al.* 1985). These investigations clearly demonstrate the difficulties involved when performing physiological studies on fish in the field: not only are standardised procedures required for capture, handling and sampling, but, in addition, the reference and polluted localities must be as similar as possible with respect to biotic and abiotic factors. Moreover, only well characterised species of fish should be used.

Although the standardised physiological methods have provided us with a large data base, comprehensive baseline data describing normal levels and the range of responses to pollutants for the parameters studied are still limited. Thus, appropriate control or reference groups must always be used for comparison. This was illustrated in a laboratory study concerning the influence of oestradiol on the induction of hepatic mixed function oxidase (MFO) by polychlorinated biphenyls (Förlin, Andersson, Koivusaari and Hansson 1983). This steroid depressed the induction of MFO activities but, since the oestrogen decreased the MFO activities in control fish as well — apparently due to an oestradiol-related inhibition of certain form(s) of cytochrome P-450 (Stegeman, Pajor and Thomas 1982) — the degree of induction was not changed.

The important role of reference groups was also illustrated in a field study by Sjöbeck, Haux, Larsson and Lithner (1984a) on feral fish from two metal polluted lakes and one reference lake. The activity of the enzyme ALA-D (see below) in red blood cells decreased steadily between August and June in fish from all three lakes. The ALA-D activity was, however, always two to three times lower in the metal-polluted fish. Thus, in spite of the seasonally variable enzyme activities, the results suggest that the levels of metal (lead) in the fish living in the polluted lakes (Sjöbeck *et al.* 1984a) did have a pronounced effect.

The use of the same response parameters in both the laboratory and field is a powerful approach for establishing sublethal toxic effects of pollutants in feral fish populations. Biochemical and physiological methods used to study the effects of cadmium, lead and pulp bleachery effluents on fish are presented and discussed below.

ALA-D Activity

The enzyme δ-aminolevulinic acid dehydratase (ALA-D) catalyses the formation of porphobilinogen and thus participates in the synthesis of haemoglobin, cytochrome and peroxidase. The activity of this enzyme in fish erythrocytes is easily measured both in laboratory and field investigations (Johansson-Sjöbeck and Larsson 1979; Hodson, Beverly and Whittle 1984; Haux *et al.* 1986). ALA-D activity is used as a sublethal response parameter because it is strongly inhibited by lead (Hodson, Blunt, Spry and Austen 1977; Johansson-Sjöbeck and Larsson 1979). This effect is a very selective and sensitive response, and probably the earliest measurable toxic response to lead exposure both in the laboratory and in the field (Table 10.2) (Larsson *et al.* 1985). In a field experiment on perch caught in waters contaminated with a complex mixture of metals, the markedly reduced ALA-D activities observed were interpreted as indicating the presence of lead in the water (Larsson *et al.* 1985). However, it should be stressed that the physiological and ecological significance of this lead-induced inhibition is uncertain. Lead may inhibit erythropoiesis; however, no obvious sign of anaemia or other haematological disturbances have been detected in lead-poisoned fish (Johansson-Sjöbeck and Larsson 1979; Haux *et al.* 1986).

Carbohydrate Metabolism

The most thoroughly studied stress responses in fish are the typical increases in blood glucose and lactate, and the depletion of glycogen in the liver and muscle. These responses are secondary effects caused by the increased release of hormones from the pituitary, the inter-renals, and the chromaffin tissues, and the increased activity of the sympathetic nervous system (Mazeaud and Mazeaud 1981). The secondary stress responses are always monitored in our field experiments in order to ensure a proper experimental performance.

In our own laboratory experiments, sublethal effects on carbohydrate metabolism have been induced by many pollutants (for references see Larsson *et al.* 1985), including cadmium, lead and pulp bleachery effluents (Tables 10.3 and 10.4). In particular, the

Table 10.2: Effects of Lead on Erythrocytic ALA-D Activity in Fish

	Laboratory effects	Field effects	
		Lead-polluted lakes	Complex mixture of metals
ALA-D activity	↓↓↓	↓↓↓	↓↓

↓ = Decrease.

Table 10.3: Effects of Cadmium on Various Physiological and Biochemical Response Parameters in Fish

	Laboratory effects	Field effects
Blood glucose	↑↑	0
Blood lactate	↑↑	↑
Muscle glycogen	↓↓	(↓)
Liver glycogen	↓↓	↓
Plasma calcium	↓↓	0
Plasma magnesium	↑	↑
Haemoglobin	↓	↓
White blood cell count	↑↑	↑↑
MFO activity	↓	NM
Metallothionein	↑↑	↑↑

↑ = Increase; ↓ = decrease; 0 = no effect; NM = not measured.

Table 10.4: Effects of Pulp Effluents from Bleach Plants on Physiological and Biochemical Response Parameters in Fish

	Laboratory effects	Field effects		
		Spring	Summer	Autumn
MFO activity	↑↑	↑↑↑	↑	↑↑↑
UDPGT* activity	0	↑	0	↑
Blood glucose	0	↓	↓	0
Blood lactate	0	↑↑	0	↑
Muscle glycogen	↑	↑↑	↑	↑↑
Liver glycogen	0	↑	↑	↑
Plasma sodium	0	0	0	0
Plasma chloride	↓	↓↓	↓↓	↓↓

↑ = Increase; ↓ = decrease; 0 = no effect.
*UDPGT = uridine diphosphate glucuronyl transferase.

cadmium-induced hyperglycaemia and depletion of muscle glycogen have physiological and toxicological significance: these effects still persisted one year after exposure, whereas the effects on other parameters had disappeared (Haux and Larsson 1984). The persistent effects of cadmium may indicate an irreversible imbalance in endocrine control primarily due to an inhibitory effect of cadmium on insulin secretion in the β-cells in the pancreas. This contention is supported by evidence that cadmium is selectively accumulated in the pancreas of fish (Havu 1969).

To reveal the ecological significance of the cadmium effects, perch found in a cadmium-polluted river were studied (Sjöbeck, Haux, Larsson and Lithner 1984b). In spite of the highly elevated levels of cadmium in the fish, the effects of cadmium on carbohydrate metabolism were slight or non-existent. These findings should, however, be viewed in the light of the adaptation of fish that occurs during exposure to cadmium (Duncan and Klaverkamp 1983). Possibly, cadmium binds to metal binding proteins synthesised in the adapted fish, thereby reducing the toxic effect of this metal (Olsson and Haux, 1985) (see also the following section).

Metallothionein

Metallothionein (MT) consists of a group of cytosolic, low molecular weight proteins that bind certain metals, and has been detected and characterised in several fish species (Klaverkamp, Macdonald, Duncan and Wageman 1984). MT can function as a valuable indicator of metal pollution in fish, as recently demonstrated for copper (Roch, McCarter, Matheson, Clark and Olafson 1982; Roch and McCarter 1984a,b). However, the role, if any, of MT in the detoxification of toxic metals is unclear. In one field study, the subcellular distribution of cadmium in perch from a cadmium-polluted river was determined (Table 10.3). Cadmium was located almost exclusively in the MT fraction, and there was a high correlation between liver cadmium levels and liver MT levels. In addition, the levels of copper and zinc were not affected by the cadmium load. Thus MT apparently plays a very important role in the sequestration of cadmium.

Ion Regulation

Strict ion regulation is necessary for aquatic organisms if they are to maintain water and ion homeostasis. Disturbances in ion regulation induced by pollutants are manifested by altered plasma ion concentrations.

In many laboratory and field experiments, we have measured concentrations of the monovalent plasma ions, sodium and chloride. These are the dominant ions in blood plasma and play an important role in maintaining osmotic pressure. Several metals (Larsson *et al.* 1985), and chlorinated hydrocarbons (Förlin *et al.* 1979; Haux and Larsson 1979) are known to affect plasma ion concentrations. In addition, recent laboratory and field tests have shown that pulp effluents from bleach plants affect plasma chloride levels (Table 10.4). Thus the concentrations of monovalent ions in blood plasma are commonly altered in response to pollutants. By measuring changes in plasma ion levels, sublethal responses to pollutants can be effectively monitored: such measurements are easily made and the methods are inexpensive, fairly sensitive and provide good indices of pathological changes in ion regulatory tissues (Larsson *et al.* 1985).

Concentrations of the divalent calcium ion are typically reduced by cadmium exposure under laboratory conditions (Roch and Maly 1979; Larsson *et al.* 1981) (Table 10.3). By monitoring both the hypocalcaemia response and carbohydrate metabolism (see above: 'Carbohydrate metabolism'), small sublethal physiological effects of cadmium in fish can be detected. The hypocalcaemia mainly affects the free plasma fraction in normocalcaemic fish due to the reduced tubular resorption of calcium in the kidney, while additional reductions in total plasma calcium in vitellogenic females can be related both to a decrease of calcium bound to vitellogenin and to a decrease in the amount of plasma vitellogenin (Larsson *et al.* 1981; Haux 1985).

Mixed Function Oxidase System

The excretion of lipid-soluble organic compounds in fish and other vertebrates is facilitated by enzyme systems converting the compounds to more polar products (Bend and James 1978). The initial biotransformation of the lipophilic compound is usually associated with an enzyme system known as mixed function oxidase system (MFO), cytochrome P-450 monooxygenase system or drug metabolising system. The MFO system introduces functional groups into the substrate which facilitates the further biotransformation by conjugating enzymes. MFO activities can be induced in response to certain lipophilic aromatic hydrocarbons, e.g. polycyclic aromatic hydrocarbons, and polychlorinated and polybrominated biphenyls (PAH-type inducers), which have resulted in over 100-fold elevations of certain MFO activities in the laboratory. The study of MFO induction has received a great deal of attention; for example, the inducing properties of aromatic hydrocarbons have been investigated (Stegeman 1981) and the use of the induction response in biological monitoring programmes has been reviewed (Payne 1984). The induction of MFO is a very sensitive and selective response towards PAH-type inducers.

In laboratory studies, fish exposed to pulp effluents from bleach plants showed markedly increased hepatic microsomal ethoxy-resorufin-O-deethylase (EROD, an MFO enzyme) activity (Förlin *et al.* 1985). Fish caught in the receiving body of water below a pulp bleach plant also had markedly increased hepatic EROD activities (5 to 20-fold) (Förlin *et al.* 1985) and this effect was observed at three different times of the year (Table 10.4). These results strongly suggest that PAH-type inducers were present in the effluent from the pulp mill, and show the considerable diagnostic value of this response parameter. Unfortunately, information regarding the toxicological and physiological significance of an induced MFO in response to PAH is very limited.

Conclusion

Methods have been designed and successfully used to monitor the effects of pollutants on fish. Most of the techniques are based on previously existing knowledge contributed by fish physiologists. The measurement of sublethal responses facilitates the early detection of pollutants present at low levels and also offers insight into the mode of action of the toxic compounds. Field investigations require standardised capture, handling and sampling procedures. Variations in the response parameters due to biotic and abiotic factors must always be considered since they can lead to misinterpretation of results. The parallel use of the same response parameters in laboratory and field studies aids in interpreting data collected from fish caught in polluted lakes, rivers and coastal areas.

References

Andersson, T. (1985) 'Regulation of Xenobiotic Metabolism in Rainbow Trout: Effects of Induction, Ambient Temperature and Starvation', thesis, University of Göteborg, Göteborg

Bend, J.R. and James, M.O. (1978) 'Xenobiotic Metabolism in Marine and Freshwater Species', in D.C. Malin and J.R. Sargent (eds), *Biochemical and Biophysical Perspectives in Marine Biology*, vol. 4, Academic Press, New York

Björnsson, B.Th., Haux, C., Förlin, L. and Deftos, L.J. (1986) 'The Involvement of Calcitonin in the Reproduction Physiology of the Rainbow Trout', *Journal of Endocrinology, 108*, 17-23

Dave, G., Johansson-Sjöbeck, M.-L., Larsson, Å., Lewander, K. and Lidman, U. (1975) 'Metabolic and Hematological Effects of Starvation in the European Eel, *Anguilla anguilla L.*: I. Carbohydrate, Lipid, Protein and Inorganic Ion

Metabolism', *Comparative Biochemistry and Physiology, 52A*, 423-30

Duncan, D.A. and Klaverkamp, J.F. (1983) 'Tolerance and Resistance to Cadmium in White Suckers (*Catostomus commersoni*) Previously Exposed to Cadmium, Mercury, Zinc or Selenium', *Canadian Journal of Aquatic Sciences, 40*, 128-38

Förlin, L. and Andersson, T. (1985) 'Storage Conditions of Rainbow Trout Liver Cytochrome P-450 and Conjugating Enzymes', *Comparative Biochemistry and Physiology, 80B*, 569-72

Förlin, L. and Hansson, T. (1982a) 'Effects of Treated Municipal Wastewater on the Hepatic, Xenobiotic, and Steroid Metabolism in Trout Liver', *Ecotoxicology and Environmental Safety, 6*, 41-8

Förlin, L. and Hansson, T. (1982b) 'Effects of Oestradiol-17β and Hypophysectomy on Hepatic Mixed Function Oxidases in Rainbow Trout', *Journal of Endocrinology, 95*, 245-52

Förlin, L., Hansson, T., Haux, C., Johansson-Sjöbeck, M.-L., Larsson, Å. and Lidman, U. (1979) 'Subletala fysiologiska effekter av PCB och DDT på skrubbskädda (*Platichthys flesus*) i bräckt och marin miljö', NBL Rapp 88, Brackish Water Toxicology Laboratory, Swedish Environment Protection Board, Stockholm

Förlin, L., Andersson, T., Koivusaari, U. and Hansson, T. (1983) 'Influence of Biological and Environmental Factors on Hepatic Steroid and Xenobiotic Metabolism in Fish: Interaction with PCB and β-Naphthoflavone', *Marine Environmental Research, 14*, 47-58

Förlin, L., Andersson, T., Bengtsson, B.-E. Härdig, J. and Larsson, Å. (1986) 'Effects of Pulp Bleach Plant Effluents on Hepatic Xenobiotic Biotransformation Enzymes in Fish: Laboratory and Field Studies', *Marine Environmental Research, 17*, 109-12

Hansson, T., Förlin, L., Rafter, J. and Gustafsson, J.-Å. (1982) 'Regulation of Hepatic Steroid and Xenobiotic Metabolism in Fish', in E. Hietanen, M. Laitinen and O. Hänninen (eds), *Cytochrome P-450 Biochemistry, Biophysics and Environmental Implications*, Elsevier, Amsterdam

Haux, C. (1985) 'Aspects of Ion Balance and Metabolism in Teleost Fish in Relation to Vitellogenin Synthesis and during Cadmium Exposure', thesis, University of Göteborg, Göteborg

Haux, C. and Larsson, Å. (1979) 'Effects of DDT on Blood Plasma Electrolytes in the Flounder, *Platichthys flesus* L., in Hypotonic Brackish Water', *Ambio, 8*, 171-3

Haux, C. and Larsson, Å. (1982) 'Influence of Inorganic Lead on the Biochemical Blood Composition in the Rainbow Trout, *Salmo gairdneri*', *Ecotoxicology and Environmental Safety, 6*, 28-34

Haux, C. and Larsson, Å. (1984) 'Long-term Sublethal Physiological Effects on Rainbow Trout, *Salmo gairdneri*, during Exposure to Cadmium and after Subsequent Recovery', *Aquatic Toxicology, 5*, 129-42

Haux, C., Larsson, Å., Lidman U., Förlin, L., Hansson, T. and Johansson-Sjöbeck, M.-L. (1982) 'Sublethal Physiological Effects of Chlorinated Paraffins on the Flounder, *Platichthys flesus* L.', *Ecotoxicology and Environmental Safety, 6*, 49-59

Haux, C., Larsson, Å., and Sjöbeck, M.-L. (1985) 'Physiological Stress Responses in a Wild Fish Population of Perch (*Perca fluviatilis*) after Capture and during Subsequent Recovery', *Marine Environmental Research, 15*, 77-95

Haux, C., Larsson, Å., Lithner, G. and Sjöbeck, M.L. (1986) 'A Field Study of Physiological Effects on Fish in Lead-contaminated Lakes', *Environmental Toxicology and Chemistry*, in press

Havu, N. (1969) 'Sulfhydryl Inhibitors and Pancreatic Islet Tissue', *Acta*

Endocrinologica Supplement, *139*, 1-231

Hodson, P.V., Blunt, B.R., Spry, D.J. and Austen, K. (1977) 'Evaluation of Erythrocyte δ-Aminolevulinic Acid Dehydratase Activity as a Short-term Indicator in Fish of Harmful Exposure to Lead', *Journal of the Fisheries Research Board of Canada, 34*, 501-8

Hodson, P.V., Beverly, R.B. and Whittle, D.M. (1984) 'Monitoring Lead Exposure of Fish', in V.W. Cairns, P.V. Hodson and J.O. Nriagu (eds), *Contaminant Effects on Fisheries*, Wiley, New York

Johansson-Sjöbeck, M.-L. and Larsson, Å. (1978) 'The Effect of Cadmium on the Hematology and on the Activity of delta-Aminolevulinic Acid Dehydratase (ALA-D) in Blood and Hematopoietic Tissues of the Flounder, *Pleuronectus flesus* L.', *Environmental Research, 17*, 191-204

Johansson-Sjöbeck, M.-L. and Larsson, Å. (1979) 'Effects of Inorganic Lead on Delta-aminolevulinic Acid Dehydratase Activity and Hematological Variables in the rainbow trout, *Salmo gairdneri*', *Archives of Environmental Contamination and Toxicology, 8*, 419-31

Klaverkamp, J.F., Macdonald, W.A., Duncan, D.A. and Wageman, R. (1984) 'Metallothionein and Acclimation to Heavy Metals in Fish: a Review', in V.W. Cairns, P.V. Hodson and J.O. Nriagu (eds), *Contaminant Effects on Fisheries*, Wiley, New York

Larsson, Å. (1975) 'Some Biochemical Effects of Cadmium on Fish', in J.H. Koeman and J.J.T.W.A. Strik (eds), *Sublethal Effects of Toxic Chemicals on Aquatic Animals*, Elsevier, Amsterdam

Larsson, Å. and Haux, C. (1982) 'Altered Carbohydrate Metabolism in Fish Exposed to Sublethal Levels of Cadmium', *Journal of Environmental Biology, 3*, 71-81

Larsson, Å., Bengtsson, B.-E. and Haux, C. (1981) 'Disturbed Ion Balance in Flounder, *Platichthys flesus* L. Exposed to Sublethal Levels of Cadmium', *Aquatic Toxicology, 1*, 19-35

Larsson, Å., Haux, C. and Sjöbeck, M.-L. (1984) 'Field Application of Physiological Methods on Fish from Metal-polluted Waters', in G. Personne, E. Jasper and C. Claus (eds), *Ecotoxicological Testing for the Marine Environment*, vol. 2, State University of Ghent and Institute for Marine Scientific Research, Bredene, Belgium

Larsson, Å., Haux, C. and Sjöbeck, M.-L. (1985) 'Fish Physiology and Metal Pollution: Results and Experiences from Laboratory and Field Studies', *Ecotoxicology and Environmental Safety, 9*, 250-81

Lehtinen, K.-J., Larsson, Å. and Klingstedt, G. (1984) 'Physiological Disturbances in Rainbow Trout, *Salmo gairdneri* (R.), Exposed at Two Temperatures to Effluents from a Titanium Dioxide Industry', *Aquatic Toxicology, 5*, 155-66

Mazeaud, M.M. and Mazeaud, F. (1981) 'Adrenergic Responses to Stress in Fish', in A.D. Pickering (ed.), *Stress in Fish*, Academic Press, New York

Olsson, P.-E. and Haux, C. (1986) 'Alterations in Hepatic Metallothionein Content in Perch, Environmentally Exposed to Cadmium', *Marine Environmental Research, 17*, 181-3

Payne, J.F. (1984) 'Mixed-function Oxygenase in Biological Monitoring Programs: Review of Potential Usage in Different Phyla of Aquatic Animals', in G. Personne, E. Jasper and C. Claus (eds), *Ecotoxicological Testing for the Marine Environment*, vol. 1, State University of Ghent and Institute for Marine Scientific Research, Bredene, Belgium

Roch, M. and Maly, E.J. (1979) 'Relationship of Cadmium-induced Hypocalcemia with Mortality in Rainbow Trout (*Salmo gairdneri*) and the Influence of Temperature on Toxicity', *Journal of the Fisheries Research Board of Canada, 36*, 1297-1303

Roch, M. and McCarter, J.A. (1984a) 'Hepatic Metallothionein Production and Resistance to Heavy Metals by Rainbow Trout. I. Exposed to an Artificial Mixture of Zinc, Copper and Cadmium', *Comparative Biochemistry and Physiology, 74C,* 71-6

Roch, M. and McCarter, J.A. (1984b) 'Hepatic Metallothionein Production and Resistance to Heavy Metals by Rainbow Trout. II. Held in a Series of Contaminated Lakes', *Comparative Biochemistry and Physiology, 74C,* 77-82

Roch, M., McCarter, J.A., Matheson, A.T., Clark, M.J.R. and Olafson, R.W. (1982) 'Hepatic Metallothionein in Rainbow Trout (*Salmo gairdneri*) as an Indicator of Metal Pollution in the Campbell River System', *Canadian Journal of Fishery and Aquatic Sciences, 39,* 1596-1601

Sjöbeck, M.-L., Haux, C., Larsson, Å. and Lithner, G. (1984a) 'Seasonal Variations in Physiological Parameters in Perch, *Perca fluviatilis,* Living in Heavy Metal Contaminated and Reference Lakes in Northern Sweden', NBL Rapp 148, Brackish Water Toxicology Laboratory, Swedish Environment Protection Board, Stockholm

Sjöbeck, M.-L., Haux, C., Larsson, Å. and Lithner, G. (1984b) 'Biochemical and Hematological Studies on Perch, *Perca fluviatilis,* from the Cadmium Contaminated River Emån', *Ecotoxicology and Environmental Safety, 8,* 303-12

Stegeman, J.J. (1981) 'Polynuclear Aromatic Hydrocarbons and their Metabolism in the Marine Environment', in H.V. Gelboin and P.O.P. Ts'o (eds), *Polycyclic Hydrocarbons and Cancer,* vol. 3, Academic Press, New York

Stegeman, J.J., Pajor, A.M. and Thomas, P. (1982) 'Influence of Estradiol and Testosterone on Cytochrome P-450 and Monooxygenase Activity in Immature Brook Trout, *Salvelinus fontinalis*', *Biochemical Pharmacology, 31,* 3979-89

11 TOXICITY TESTING PROCEDURES

Göran Dave

From a regulatory point of view, toxicity tests are used for three major purposes. These are (1) screening of chemicals and products, (2) establishing limits, and (3) monitoring (Cairns 1984). For screening and monitoring, inexpensive and rapid tests with known and preferably high precision are needed. For establishment of limits, the tests must not necessarily be inexpensive and rapid, but the effects measured should be directly or causally related to environmental hazard, easily interpreted, and meaningful to the public and courts.

In his already classic textbook *Biology and Water Pollution Control*, Warren (1971) emphasised the importance of studying environmental problems at all levels of biological organisation, from the molecular up to the ecosystem level. Studies on lower levels of organisation can help us understand the mechanisms behind environmental changes. But our knowledge of the function of biological systems and in particular their response to chemicals is insufficient for detailed predictions from lower to higher levels of organisation. Therefore, the significance of a response observed at a low level can only be revealed by studies at higher levels of organisation.

It may be debated which level to start the study of the environmental hazard of a new chemical. Should it be at the molecular or ecosystem level? I will leave this question unanswered, to be kept in the reader's mind while reading this chapter. Whatever the final answer will be, a basic understanding of the advantages, disadvantages and limitations of the various approaches used is essential (Cairns 1980, 1983).

Traditionally, single, surrogate-species toxicity tests have been used to establish limits for protection of human health and ecosystem integrity, and this presentation is traditional in dealing almost exclusively with results and conclusions from single-species toxicity tests. However, the examples have been organised with the intention of giving some insight into what may occur at the next highest level of organisation, the population living under variable conditions, here called *abiotic factors*. In addition attention has

170

been drawn to some intrinsic properties of living organisms of known or suspected importance for their responses to toxicants, here called *biotic factors*. Knowing the interactions caused by the abiotic and biotic factors, we are in a better position both to predict and to prevent toxicity. But first of all the three major response variables in toxicity tests with fish (mortality, development and reproduction) will be discussed briefly.

The references cited are, because of the extent of the literature, only examples. They are given to provide the reader with details on practical and theoretical aspects. In addition they will make possible a critical examination of my thoughts and conclusions.

Response Variables

Any variable exhibiting a dose-dependent response may be used in a toxicity test. For practical use, easily quantified variables are preferable; the easier to measure the better the precision. However, also easily measured variables, like counting and sizing of algae, daphnids and fish, need some experience to be determined accurately, especially when done repeatedly in a test in order to minimise stress due to handling.

Mortality

Mortality is different from most sublethal variables because it is a quantal, i.e. an all-or-none, response. It may be replaced by immobility in tests with small animals and in tests with anaesthetics. Quantal responses are usually analysed by the log-probit model to derive EC or LC50 values. Because of its common use in aquatic toxicology, data from acute toxicity tests are generally treated by statistical programs to determine not only the LC50, but also its 95 per cent confidence or fiducial limits, the slope of the dose-response relationship, its variance and the χ^2 for fitness of model. The importance of the experimental design, like the number of fish per concentration, number of concentrations, dilution factor, and number of replicates, is easy to overlook if computers are used as magic black boxes for determination of these accessory LC50 figures. In order to avoid unnecessary work and false conclusions in acute toxicity testing, the papers by Hodson, Ross, Niimi and Spry (1977) and Stephan (1977) are recommended for study.

Practical considerations in toxicity testing with fish and other aquatic animals are provided by APHA (1985), the Committee on Methods for Toxicity Tests with Aquatic Organisms (1975), ISO (1982, 1984) and Scherer (1979).

Development and Growth

The terminology of teleost development is poorly standardised, with green eggs, eyed eggs, fry, and pre- and post-larvae used to characterise various stages of research animals. According to Balon (1975), the entire life of a fish can be separated into five periods: the embryonic, larval, juvenile, adult and senescent periods. The embryonic period can be divided further into three phases (the cleavage, embryonic and eleutheroembryonic phases), and the larval period into another two phases (protopterygiolarval and pterygiolarval). In salmonids the juvenile period can be divided into three phases (alevin, smolt and juvenile). However, unlike the situation in many invertebrates, the developmental periods and phases of teleosts have no distinct bounds and are not suited for quantal response models. The only exception is the transition from the embryonic to the eleutheroembryonic phase, the hatching. Therefore the timing of hatching is the most accurate measurement to be used for determination of effects on early development, and it is somewhat surprising that this possibility has been so little explored until quite recently. Examples of chemicals that have been shown to affect hatching time are shown in Table 11.1. Under sublethal conditions of exposure, hatching may be either inhibited (delayed) or stimulated (enhanced), and both effects may occur with the same chemical, as seen with MS-222, revealing the importance of the dose. A tentative working hypothesis (Dave 1985) is that a slight stress can stimulate hatch, whereas a severe stress can inhibit hatch and ultimately prevent it completely by killing the embryo.

The ecological significance of an inhibited or a stimulated hatch may be disputed (Balon 1984), but this is not uncommon for sublethal effects as mentioned in the introduction. Rosenthal and Alderdice (1976) in their extensive review suggested how primary ('hidden') effects on early developmental stages may show up as secondary quantifiable effects and what the consequences may be later on (tertiary effects).

Instead of focusing on the imprecise early developmental stages, growth (length and/or weight) has become the sublethal variable

Table 11.1: Examples of Toxicants Affecting Time to Hatching in Fish

Toxicant	Species	Observed effect on hatching	Reference
Aliguat 336	*Brachydanio rerio*	Stimulation	Dave *et al.* 1981
Aluminium	*Brachydanio rerio*	Stimulation at high pH	Dave 1985
Cadmium	*Brachydanio rerio*	Stimulation at high pH	Dave 1985
Chromium (II)	*Salmo gairdneri*	Inhibition	Stevens and Chapman 1984
Chromium (VI)	*Salmo salar*	Stimulation	Grande and Andersen 1983
DDT	*Oryzias latipes*	Stimulation and inhibition	Takimoto *et al.* 1984
Fenitrothion	*Oryzias latipes*	Inhibition	Takimoto *et al.* 1984
Iron (III)	*B. rerio*	Stimulation at neutral pH	Dave 1985
Lead (II)	*S. salar*	Stimulation	Grande and Andersen 1983
Mercury (HgCl$_2$)	*S. gairdneri*	Stimulation	Klaverkamp *et al.* 1983
MS-222	*Fundulus heteroclitus*	Stimulation and inhibition	DiMichele and Taylor 1981
Nickel	*S. salar*	Inhibition	Grande and Andersen 1983
PCP	*S. gairdneri*	Inhibition	Dominguez and Chapman 1984
pH (4.2-8.5)	*B. rerio*	Maximal stimulation at pH 6.2	Dave 1985
pH (4.0-6.4)	*Perca fluviatilis*	Inhibition at low pH	Rask 1984
TFM	*S. gairdneri*	Inhibition	Niblett and McKeown 1980
Vanadium	*S. gairdneri*	Stimulation	Giles and Klaverkamp 1982
Zinc	*S. salar*	Inhibition	Grande 1967
Zinc	*S. trutta*	Inhibition	Grande 1967
Zinc	*Pimephales promelas*	Inhibition	Brungs 1969

of choice in most EL (embryo-larval) toxicity tests with fish. The philosophy of using this response is that many different non-detected effects of toxicants may show up as an impaired ability to convert food into growth. Woltering (1984) reviewed its utility in 173 chronic and EL toxicity tests. He concluded that survival alone would be sufficient for routine EL toxicity testing with fish. Arguments for omitting the growth response were its questionable ecological significance (sublethal response), its transient nature found early in tests with several toxicants, the occurrence of toxicant-induced stimulation (=hormesis; reviewed by Stebbing 1981, 1982), and the influence of ration and individual interaction during group-rearing as affected by partial kills. Another important point is individual variability in growth, which is probably most affected by food particle size and composition. These two qualities are probable reasons for the superiority of using live food in EL toxicity tests and larval rearing of 'difficult' cultured aquatic animals — 'the brine shrimp success'. Furthermore, the movement, taste, digestibility and biochemical composition of live foods may be more important than we generally assume (Dabrowski 1984).

One question which has not been addressed systematically in EL toxicity tests is the effect of feeding level on the sensitivity. Tests with *Daphnia* have shown that the feeding level can affect the sensitivity to toxicants in both directions (Lewis and Weber 1985; Taylor 1985). An intermediate ration would reduce such effects and maybe improve the sensitivity of the growth response, but probably at the expense of a higher variability. Such interactions and the variability among larvae of different fish species in food particle density and size dependency (Theilacker and Dorsey 1980) point to the need for more basic research on our test animals as proposed by Maki (1983). However, I do not agree with the view that efforts towards standardisation of test methods are doomed to fail until we have more knowledge of these basic requirements. On the contrary, exercises called intercalibrations, ring tests or round robins (How 1980) are instrumental in critically examining repeatability and reproducibility of test methods used as regulatory tools.

Reproduction

Reproduction is generally considered to be a delicate process sensitive to the effects of toxicants and other stressors (Donaldson and Scherer 1983). However, experimentally it is probably the most difficult process to study, and the chronic life-cycle tests with fish conducted during the last 20 years are accomplishments worth all respect.

Desirable features of fish species used for chronic tests are: relative easiness of laboratory rearing, short life cycle, small size and non-adhesiveness of eggs. In addition, possibilities for paired spawning are desirable, since group spawning restricts the analysis of reproductive performance and may result in egg predation. But there will still be a variation in day-to-day egg production. Gerking (1980) has reviewed these problems in more detail. The desert pupfish (*Cyprinodon n. nevadiensis*) used in his laboratory seems to be an almost ideal species for studies on reproductive performance. But in spite of efforts to reduce intrinsic variability, the CV (coefficient of variation) for day-to-day egg production during a 21-day period has been around 100 per cent, and the CV between females around 60 per cent.

Among the test results reviewed by Woltering (1984), reproduction was the single most sensitive response in 15 per cent of the chronic tests and equally sensitive with one or more other responses in another 15 per cent. Toxicants having reproduction (fecundity) as the single most sensitive response included several metals (Cd, Cu, Pb, Zn). However, the pronounced effect of zinc on egg production found in the fathead minnow by Brungs (1969) was later on revealed to be due to a decreased adhesiveness and chorion strength induced by early zinc exposure, and not to fecundity. This peculiar effect of zinc was revealed by the use of an egg-trap under the spawning tile (Benoit and Holcombe 1978). This is a good example of the importance of studying effects of pollution at several levels of biological organisation.

The studies by Gerking and co-workers just cited, on the effects of pH, salinity and temperature in the desert pupfish, showed that criteria for reproductive performance must include both egg production and egg viability. Spawning in water which is toxic to the eggs or larvae is not uncommon. Because of the high sensitivity of early developmental stages, exposure of only embryos and larvae has been of equal sensitivity to exposure including adults and spawning in the majority of chronic tests. Only rarely has reproduction (fecundity) been more than twice as sensitive as larval survival and growth (McKim 1977). Therefore, the EL toxicity test has become a cost-efficient substitute for the chronic test. The chronic test is still however, the base from which to judge physiological responses and such tests are necessary to keep to the main scientific course while developing short cuts (Mount 1977).

Influence of Abiotic Factors

Ecological effects of pollutants can be primary, secondary and perhaps also tertiary (Hurlbert 1975). One example of a secondary effect is the abundance of invertebrate predators in severely acidified lakes, which is due to loss of fish (Henriksson and

Oscarsson 1981). This is an effect at biological interaction level which is impossible to predict quantitatively, and which can illustrate our limited capability to predict secondary effects. However, there are also effects of toxicants that are directly affected by natural variability which can be mimicked in laboratory tests. These effects can act outside (abiotic) or inside (biotic) the fish. They are important to consider when laboratory and field data are to be compared.

Alkalinity, Hardness and pH

Alkalinity and pH are generally correlated with hardness, low values occurring in soft water and high values in hard water. Numerous studies have shown that the acute toxicity of cadmium, copper and zinc is higher in soft compared with hard water, when expressed as the total concentrations of the metals. But when expressed as the concentrations of the free metals, the effect of pH appears to be the opposite, i.e. these metal ions are more toxic at high compared with low pH. However, the effect of hardness (calcium and magnesium) can obscure this effect of pH (Borgmann 1983). Detergents comprise another group of chemicals whose toxicity is known to be affected by hardness (Tovell, Newsome and Howes 1974; Maki and Bishop 1979; Lewis and Perry 1981).

Among all abiotic factors affecting toxicity to aquatic organisms, pH is the most powerful one. Within its non-lethal limits of 5 and 9 for fish (Alabaster and Lloyd 1982) the toxicity of weak acids and bases may be affected by orders of magnitude. Well-known examples are ammonia, which is more toxic at high pH (Alabaster and Lloyd 1982), and chlorinated phenols, which are generally more toxic at low pH (Könemann and Musch 1981). Pharmacokinetic models may predict this interaction (Hayton and Stehly 1983), but more recent studies have shown that the ionised forms of the chemicals mentioned are also toxic, especially during long-term exposure (Dave 1984; Broderius, Drummond, Fiandt and Russom 1985; Spehar, Nelson, Swanson and Renoos 1985). Other common pollutants with pH-dependent toxicity are cyanide, nitrite and sulfide (Broderius, Smith and Lind 1977; Russo, Thurston and Emerson 1981), and some fishery chemicals are also strongly affected by pH, e.g. antimycin (Marking 1975), MS-222 (Ohr 1976), quinaldine sulfate (Marking and Dawson 1973) and TFM (Marking and Olson 1975).

Dissolved Organics

The effect of dissolved organics on the toxicity of metals, and in particular on that of copper (McCrady and Chapman 1979), is well known. The mechanism of detoxification is probably analogous to that of carbonates, hydroxides and sulphates in hard water, i.e. binding of the free metal. Examples of dissolved organics with this capacity include several amino acids, buffers and complexing agents in artificial media such as citrate and EDTA, water softeners such as NTA, and unidentified excretory products of algae (Van den Berg, Wong and Chau 1979) and daphnids (Fish and Morel 1983). In addition to these organics, the fulvic and humic acids (Landrum, Reinhold Nihart and Eadie 1985) may be of particular importance in certain brown-water rivers and lakes. Because of the low concentrations of dissolved organics in most natural waters, they are probably of greater importance for chronic than for acute toxicity (Borgmann 1983). This is the major reason for the lack of studies into the effect in relation to natural concentrations of dissolved organics. Furthermore, the interpretation of their role as toxicity modulators must take into account their possible degradation, pH-dependence of complexing capacity, and seasonal variation. Sensitive test methods, multifactorial experiments and experiments with natural waters with different characteristics may reveal some of their importance as toxicity modulators.

Dissolved Oxygen

The direct effects of dissolved oxygen (DO) on fish and other aquatic organisms are beyond the scope of this presentation. Excellent reviews have been made by, for example, Davis (1975) and Alabaster and Lloyd (1982). For the majority of toxicants studied, the LC50 at a DO concentration of $5\,mg\,litre^{-1}$ has been found to be approximately half of that found at $10\,mg\,litre^{-1}$. This is because respiratory activity is often increased as a compensatory response to a decreased DO concentration, thus increasing the rate of uptake of toxicants across the gills (Lloyd 1961).

Salinity

The effect of salinity on toxicity includes several factors. One is pH, which on an average is one unit higher in sea water than in fresh water (Warren 1971). A second is mono- and divalent ion interaction on osmoregulatory physiology. A third is the different assembly of species inhabiting fresh water compared with sea water. Therefore, differences in sensitivity to pollutants in fresh water and sea water are difficult to interpret, but at least some influence mechanistically similar to that for pH and hardness should be expected.

Areas of special concern are estuaries and brackish-water areas such as the Baltic and also the very soft waters of tributaries, rivers and lakes in granite bedrock coastal areas. Such areas are common in, for example, Scandinavia, and the effects of acid rain and the resulting increases in metal concentrations may be affected by even small amounts of sea-derived ions in such areas (Brown 1981; Brown and Lynan 1981; McDonald and Wood 1982; McWilliams 1982).

Suspended Solids

Alabaster and Lloyd (1982) distinguished five ways that an excessive concentration of finely divided solids might be harmful to a fishery in a river or lake:

(1) by direct lethal or sublethal action including a reduced resistance to disease,
(2) by preventing successful development of eggs and larvae,
(3) by modifying natural movements and migrations,
(4) by reducing food abundance and availability,
(5) by affecting the efficiency of catching methods.

There was no evidence that concentrations of suspended solids less than 25 mglitre^{-1} would harm the fishery, and only marginal effects would be expected between 25 and 80 mglitre^{-1}

However, the presence of suspended solids may also affect the toxicity and uptake of chemicals, as seen in daphnids (Maki and Bishop 1979; McCarthy 1983). How the toxicity to fish is affected by suspended solids is essentially unknown. Brungs and Bailey (1966) studied its effect on the acute toxicity of endrin in the fathead minnow. Only with suspended activated carbon (12 mglitre^{-1}) was toxicity reduced. The nominal 96-h LC50 was 40 times higher with suspended activated carbon, but when expressed as dissolved endrin the 96-h LC50 values were almost identical. Lewis and Wee (1983) studied the bioconcentration and toxicity of dialkyl dimethyl ammonium surfactants in reconstituted water, well water and river water. Both bioconcentration and toxicity in algae, daphnids and fish were less in river compared with reconstituted water. Low solubility in water, strong adsorption to solids and tendency to complex with dissolved anionic compounds were concluded to be key factors for the reductions in bioconcentration and toxicity found in river water.

Apparently suspended solids can modify the toxicity of certain chemicals, but the magnitude and species dependency of this interaction certainly needs to be studied more. Such studies should preferably also incorporate sediment toxicity, for which toxicity tests have been developed (Prater and Andersson 1977; Maleug, Schuytema, Gakstatter and Krawczyk 1984; Nebeker, Cairns, Gakstatter, Maleug, Schuytema and Krawczyk 1984; LeBlanc and Surprenant 1985). If suspended solids decrease toxicity to pelagic species by adsorption, then exposure to benthic species may be increased.

Temperature

The effects of temperature on toxicity of chemicals to aquatic organisms were reviewed by Cairns, Heath and Parker (1975), and to my knowledge there is still no way to predict the effect of temperature on toxicity. An increase of 10°C may increase the acute toxicity to fish eight times as in the case of guthion, or decrease it two times as in the case of DDT (Peterson, 1976). Furthermore, the pattern of interaction may differ among chemicals, some being least toxic at intermediate temperatures (Dave, Andersson, Berglind and Hasselrot 1981), and temperature acclimation may also interact (Hodson and Sprague 1975; Kovacs and Leduc 1982).

Because of the non-predictable effect of temperature on toxicity, the need for simplified methods which can be applied systematically on a great number of chemicals is obvious. Since temperature and toxicity have much in common, models for temperature tolerance may be applicable. A recent model presented by Kilgour, McCauley and Kwain (1985) was suitable for predicting not only the incipient temperature threshold but also the threshold for copper ion toxicity from short-term tests. By using the median effective times from short-term tests, long-term effects may be predicted by the use of suitable models. Such a possibility is especially useful in studies of multifactorial interaction.

Only for chlorine, some metals, pesticides and fishery chemicals are systematically derived data available. For the majority of chemicals comparable data are lacking. Different papers dealing with temperature-toxicity interaction have been compiled on a yearly basis in the literature reviews in the *Journal of the Water Pollution Control Federation* (6), e.g. Cravens and Harrelson (1985), but a systematic approach is still lacking.

Influence of Biotic Factors

Species

Albert (1973) distinguished three principles of selective toxicity, differences in sensitivity being due to (1) differences in internal distribution, (2) biochemical differences, and (3) cytological differences. Theoretically, in order to estimate the sensitivity of entire ecosystems it would then be necessary to determine the toxicity to each species. To do this for all chemicals is not practically, economically or ethically justifiable. The two major short cuts are to estimate the risk that aquatic environments will be contaminated (Gillett 1983), and to estimate the risk for selective toxicity (LeBlanc 1984). The minimum number of species to estimate the potential selectivity is, on a strictly mathematical basis, three. In order to maximise the capability to detect selectivity, the phylogenetic distance should also be maximised. For ecological reasons organisms representing different functional groups should also be included. For these reasons an assembly of an alga, a member of the zooplankton and a fish may be used for a preliminary hazard assessment. Whether these organisms are exposed simultaneously in the same tank or in separate tanks is a matter of convenience and precision. Ecological interaction should not be confused with selective toxicity.

Developmental Stage

Generally speaking, early and late developmental stages of an animal are more susceptible than intermediate stages, which has made it possible to generate chronic toxicity estimates by means of EL toxicity tests. Further short cuts proposed are a 96-h DNA, RNA and protein content test (Barron and Adelman 1984), a 7-day fathead minnow larval test (Norberg and Mount 1985) and a 14-day zebrafish EL toxicity test (Dave, Grande, Kristensen, Martelin, Rosander and Viktor, in preparation). All three have been developed with the philosophy of only exposing the most sensitive periods in the life cycle.

In most EL toxicity tests, exposure of the earliest stages is lacking (the free gamete, fertilisation and early cleavage which is accompanied by swelling and uptake of water into the perivitelline space). Some results (Billard and Roubaud 1985; Niblett and McKeown 1980) have indicated a high sensitivity of these early stages. This aspect as well as variability in egg quality (Craik and Harvey 1984) and in sensitivity to chemicals of eggs from different parents (Alderman 1984), and, furthermore, the influence of developmental stage (Manner and Muehlman 1975) and environmental ionic composition (Cameron and Hunter 1984) on chorionic permeability may effect results and conclusions from EL toxicity tests. However, the fact that the EL toxicity testing approach was judged from chronic tests, incorporating most of these aspects (Macek and Sleight 1977; McKim 1977), may suggest that their overall impact may be less than deduced from the results mentioned above.

Another aspect of reproductive effects was addressed by Landner, Neilson, Sörensen, Tärnholm and Viktor (1985). They simulated the situation of migrating salmonids passing a pulp-mill effluent. Only prespawning adults were exposed. Spawning and embryonic and larval development took place in clean water. Surprisingly, exposure of only the males was more detrimental than exposure of only the females. When both sexes were exposed, the effect persisted at least 42 days after exposure of adults was terminated. Adults had been exposed for only 10 days. Pre-exposure of the adults was in fact more sensitive than direct exposure of embryos and larvae.

Dietary Exposure and Nutritional Interaction

Nutritional effects on toxicity are important for two major reasons. One is that they may cause non-desirable error in laboratory tests, and the other, and perhaps more important, reason is that the nutritional status is seasonally dependent in many fish species. Contrary to what was believed in the early 1960s (Carson 1962), the dietary route seems to be of less importance than the respiratory route in aquatic animals for pesticides (Ellehausen, Guth and Esser 1980) as well as for metals (Tarifeño-Silva, Kawasaki, Yu, Gordon and Chapman 1982). Therefore, bioconcentration of organic chemicals can be roughly predicted from chemical-physical properties (Davies and Dobbs 1984).

The nutritional influence on fish toxicity data was reviewed by Mehrle, Mayer and Johnson (1977), pointing to a six-fold effect on acute toxicity of chlordane. However, a more recent study by Marking, Bills and Crowther (1984) with five diets and eleven chemicals did not reveal any effects of diet. In another study (Dave 1981), the effect of lipid content and starvation was about two-fold with, as expected, lean fishes being more sensitive than fat fishes. Wild fathead minnows trapped in spring were as sensitive as minnows starved for several months, pointing to the importance of seasonal variation in lipids for the sensitivity to lipophilic chemicals.

Previous Exposure

The previous exposure of a population of fish can increase its apparent tolerance in at least three ways: (1) by killing the sensitive individuals (preselection), (2) by repeated partial kills for several generations (genetic selection), and (3) by activating physiological defence mechanisms (adaptation). The first two mechanisms have been studied extensively in the mosquitofish (*Gambusia affinis*) inhabiting cotton-field drainage ditches in the southern United States (reviewed by Andreasen 1985). So far, the same extraordinary resistance to pesticides has not been seen in other species of fish, although it has been seen in many target species of insects. Development of resistance, although of possible benefit to the species affected, may impose an ecological hazard because of the associated higher dietary exposure to natural predators, man included.

A similar development of genetic selection and resulting tolerance has not been demonstrated in fish after exposure to metals, but adaptation has been seen with zinc (Spehar 1976), cadmium (Birge, Benson and Black 1983) and arsenic and copper (Dixon and Sprague 1981a,b). Induction of metal-binding proteins such as metallothionein may be a key factor involved. One way to distinguish adaptation and selection in wild populations of fish is that tolerance due to adaptation is progressively lost following transfer to clean water.

Mixture Toxicity

Toxicity of mixtures (joint toxicity) can be either less than additive (antagonism), strictly additive, or more than additive (potentiation), according to the terminology proposed by Sprague (1970). More recent findings and proposed models in aquatic toxicology have been reviewed by Alabaster and Lloyd (1982). Two important questions in this expanding field of research are the minimum toxic contribution (Könemann 1981; Hermens, Leeuwangh and Musch 1984) and the contribution of chemicals in mixtures to the chronic toxicity. Alabaster and Lloyd (1982) concluded that concentration addition, the TU model (TU = toxic unit = actual concentration divided by the LC50), predicted the toxicity of mixtures of toxicants with TUs above 0.1 in sewage and industrial effluents. They also emphasised that the choice of model must depend on the type of information required. For instance, regulatory and research objectives are not always identical. But one approach does not rule out the other, and eventually different approaches may produce a similar conclusion.

Effluent Toxicity

There are several arguments for using toxicity tests with effluents as part of regulatory instruments. Four of these arguments are:

(1) The response time in a toxicity test is directly related to the acute environmental impact — the more toxic the faster the response.
(2) For effluents in which the major toxicants are unknown, the use of chemical monitoring is a poor guarantee against environmental impact.
(3) The response in a toxicity test incorporates not only the joint action of all toxicants, no matter what the correct model is, but also the abiotic and biotic factors dictated by the design.
(4) It will keep environmental toxicologists on course towards solving 'real world' problems, and it may become a platform for discussions between industry, government and university people oriented towards solving local environmental problems.

These arguments are not meant to give the illusion that toxicity tests are the solution to all environmental problems. However, there has been a tendency to continue to provide an increasing number of laboratory data with analytical-grade toxicants in carefully controlled experiments, but a reluctance to deal with complex mixtures of largely unknown composition such as effluents.

The development of site-specific water-quality criteria (Spehar and Carlson 1984; Hedtke and Arthur 1985; Stephan 1985) is a step towards incorporating local characteristics into effluent guidelines. Another approach to bridge the gap between laboratory and field studies is to use the same response variables in both studies. In this respect, biochemical and physiological properties (Mayer 1983; Larsson, Haux and Sjöbeck 1985), morphological characteristics (Bengtsson 1979; Carline and Lawal 1985), residues and key enzymes (Hodson, Blunt and Whittle 1983; Connolly 1985) are more suitable than mortality, growth and reproduction, which are difficult to determine in the field. The necessity to connect effects with residue levels was stressed by Orfila more than a hundred years ago (cf. Casarett and Bruce 1980) and, again, in the field of aquatic toxicology by Sprague (1976).

Seasonal Variability

All abiotic and biotic factors mentioned above show seasonal patterns, and, thus, a seasonal pattern in susceptibility to pollutants is to be expected. The overall effect of season is, however, hard to predict, because joint effects of various factors may behave like chemicals in a mixture with less than additive, additive or more than additive effects. Brown (1968) proposed a method to predict effluent toxicity by incorporating abiotic influence and concentration addition to toxicants, and this approach may be useful in examining seasonal variation. However, there is one fundamental difference between mixture toxicity and seasonal influence as regards regulatory control. Joint effects of chemicals in effluents are inherent properties of effluent tests, but seasonal influence is not. Furthermore, the seasonal influence cannot be regulated, only forecasted and retrospectively analysed. Because of the difficulties involved in predicting the overall seasonal influence, the question may be addressed in more general terms. Examples of questions to be considered are:

(1) Sensitivity before, during and after the reproductive season. These events are depending on species, but in temperate

regions they often coincide with early spring, late spring and late summer.

(2) Sensitivity during extremes of cold and warm weather.
(3) Sensitivity during periods of high and low productivity.
(4) Sensitivity during periods of high and low water flow.

These and other questions of seasonal character can be addressed by conventional methodology, e.g.

(a) Standardised toxicity tests with defined toxicants in dilution water sampled during selected periods in order to estimate the abiotic seasonal component (Borgmann 1981; Spehar and Carlson 1984).
(b) Standardised toxicity tests with organisms sampled during selected periods in order to estimate the biotic seasonal component (Tatem, Anderson and Neff 1976; Falk and Dunson 1977; Norton and Franklin 1980).
(c) Ambient condition tests with synchronised developmental phase and defined toxicants and time-independent response (Zitko and Carson 1977).

In addition to this, more recently developed tools such as quantitative-structure-activity relationships (Lipnick 1985), computerised models (O'Neil, Bartell and Gardner 1983), and biological monitoring (Cairns 1981) may become useful. An intensified research on causes of fish kills (Van Hoof, Van Craenenbroeck and Marivoet 1984) can help to identify critical periods in the environment, and the information from fish-farm experiences should not be overlooked (Dickson 1983; Karlsson-Norrgren, Dickson, Ljungberg and Runn 1985). The possible occurrence of daily rhythms in tolerance must also be considered (McLeay and Munro 1979; Bulger 1984).

Species-specific Water-quality Criteria — a New Challenge in Aquatic Toxicology

In hazard assessments and in the development of water-quality criteria (WQC) for aquatic life, toxicologists are brought face to face with the problem of extrapolating toxicity data from short-term tests to long-term exposure (Kenaga 1982) and from a limited number of species to the majority of species (LeBlanc 1984; Suter, Vaughan and Gardner 1983; Stephan 1985). This situation is the opposite to that in the direct protection of man, when extrapolations are made from several animal species to one species, *Homo sapiens* (Calabrese 1983). Only occasionally are environmental toxicologists extrapolating data to one particular species. The protection of rare and endangered species of birds and mammals may illustrate some of the difficulties involved (Peakall and Fox 1983). In the aquatic environment the need for a similar extrapolation between water-breathing species has not been put forward. Therefore, most aquatic toxicologists have not considered the difficulties involved. Suter *et al.* (1983) discussed some of the problems in statistical terms, and the classic paper of Brown (1968) contains several of the predictive problems involved, although the objective of the paper was environmental protection in general.

Is there a need for species specific WQC? Yes, I believe there is, not least because it would be a scientific challenge to lose the 'general' safety which is incorporated into the 'species safety' factor. The species to start with should be the cultured ones for three major reasons. The first is that those are the species we know most about. The second is that many WQC for protection of aquatic life are overstated when applied in aquaculture. Production can often be maintained at much higher concentrations (Wickins 1981). The third is that there are mutual benefits to achieve. Aquatic toxicologists, like all experimental biologists, have an interest in good-quality animals; aquaculturists have an interest in guidelines for water quality in order to provide more and better products; and in the interest of man there is an increasing need for high-quality protein.

So, how can aquatic toxicity data be improved to suit aquaculture needs? In spite of the fact that the two major response variables (survival and growth) are the same in most nutritional and toxicological studies, the emphasis on the results obtained is different. Toxicological studies are directed towards detection of minimum effect levels. Aquaculture studies are directed towards detection of limiting factors for production. These limiting factors may be of aetological, genetic, nutritional, pathological, technological or toxic nature. One improvement of toxicological studies could be to put more emphasis on quantitative dose-response relationships. A statistically significant difference does not necessarily imply an economically significant difference. A more quantitative effect-orientated approach would also facilitate comparisons between effects of natural and pollutional origin. A limiting-factor approach incorporating ecological variables, such as habitat suitability (Raleigh, Hickman, Solomon and Nelson 1984), limiting nutrients, pH, season and temperature could be instrumental in quantifying the ecological impact of pollutants.

However, there are differences in ambient conditions in aquaculture and the natural environment that must be considered. For instance the levels of ammonia, carbon dioxide, nitrite, nitrate oxygen, phosphate, and dissolved and suspended organics may be different. Nutritional factors and susceptibility to transmission of infectious diseases are also different. But interactions with these factors are also important in aquatic toxicology. A united effort towards a better understanding of the toxicity to our cultured species may become very profitable both economically and scientifically. Not least for biological monitoring may aquaculture operations have an enormous potential.

References

Alabaster, J.S. and Lloyd, R. (1982) *Water Quality Criteria for Freshwater Fish*, Butterworth Scientific, London

Albert, A. (1973) *Selective Toxicity*, fifth edn, Chapman & Hall, London

Alderman, D.J. (1984) 'The Toxicity of Iodophors to Salmonid Eggs', *Aquaculture, 40*, 7-16

Andreasen, J.K. (1985) 'Insecticide Resistance in Mosquitofish of the Lower Rio Grande Valley of Texas — an Ecological Hazard?', *Archives of Environmental Contamination and Toxicology, 14*, 573-7

APHA (1985) *Standard Methods for the Examination of Water and Wastewater*, 16th edn, American Public Health Association, Washington, DC

Balon, E.K. (1975) 'Terminology of Intervals in Fish Development', *Journal of the Fisheries Research Board of Canada, 32*, 1663-70

Balon, E.K. (1984) 'Reflections on Some Decisive Events in the Early Life of Fishes', *Transactions of the American Fisheries Society, 113*, 178-85

Barron, M.G. and Adelman, I.R. (1984) 'Nucleic Acid, Protein Content, and Growth of Larval Fish Sublethally Exposed to Various Toxicants', *Canadian Journal of Fisheries and Aquatic Sciences, 41*, 141-50

Bengtsson, B.E. (1979) 'Biological Variables, Especially Deformities in Fish, for Monitoring Marine Pollution', *Philosophical Transactions of the Royal Society of London, Series B (Biology), 286*, 457-64

Benoit, D.A. and Holcombe, G.W. (1978) 'Toxic Effects of Zinc on Fathead Minnows, *Pimephales promelas*, in Soft Water', *Journal of Fish Biology, 13*, 701-8

Billard, R. and Roubaud, P. (1985) 'The Effect of Metals and Cyanide on Fertilization in Rainbow Trout (*Salmo gairdneri*)', *Water Research, 2*, 209-14

Birge, W.J., Benson, W.H. and Black, J.A. (1983) 'The Induction of Tolerance to Heavy Metals in Natural and Laboratory Populations of Fish', Research Report no. 141, Water Resources Research Institute, University of Kentucky, Lexington, Kentucky, 26 pp.

Borgmann, U. (1981) 'Determination of Free Metal Ion Concentrations Using Bioassays', *Canadian Journal of Fisheries and Aquatic Sciences, 38*, 999-1002

Borgmann, U. (1983) 'Metal Speciation and Toxicity of Free Metal Ions to Aquatic Biota', in J.O. Nriagu (ed.), *Aquatic Toxicology, Advances in Environmental Science and Technology*, vol. 13, Wiley, New York, pp. 47-72

Broderius, S.J., Smith Jr, L.L. and Lind, D.T. (1977) 'Relative Toxicity of Free Cyanide and Dissolved Sulfide Forms to the Fathead Minnow *Pimephales promelas*', *Journal of the Fisheries Research Board of Canada, 34*, 2323-32

Broderius, S., Drummond, R., Fiandt, J. and Russom, C. (1985) 'Toxicity of Ammonia to Early Life Stages of the Smallmouth Bass at Four pH Values', *Environmental Toxicology and Chemistry, 4*, 87-96

Brown, D.J.A. (1981) 'The Effects of Various Cations on the Survival of Brown Trout, *Salmo trutta*, at Low pHs', *Journal of Fish Biology, 18*, 31-40

Brown, D.J.A. and Lynan, S. (1981) 'The Effects of Sodium and Calcium Concentrations on the Hatching of Eggs and the Survival of the Yolk Sac Fry of Brown Trout, *Salmo trutta* L., at Low pH', *Journal of Fish Biology, 19*, 205-11

Brown, V.M. (1968) 'The Calculation of the Acute Toxicity of Mixtures of Poisons to Rainbow Trout', *Water Research, 2*, 723-33

Brungs, W.A. (1969) 'Chronic Toxicity of Zinc to the Fathead Minnow, *Pimephales promelas*', *Transactions of the American Fisheries Society, 98*, 272-9

Brungs, W.A. and Bailey, G.W. (1966) 'Influence of Suspended Solids on the Acute Toxicity of Endrin to Fathead Minnows', *Proceedings of the 21st Purdue Industrial Waste Conference, Part 1, 50*, 4-12

Bulger, A.J. (1984) 'A Daily Rhythm in Heat Tolerance in the Salt Marsh Fish *Fundulus heteroclitus*', *Journal of Experimental Zoology, 230*, 11-16

Cairns, J. Jr (1980) 'Estimating Hazard', *Bioscience, 30*, 101-7

Cairns, J. Jr (1981) 'Biological Monitoring, Part VI — Future Needs', *Water Research, 15*, 941-52

Cairns, J. Jr (1983) 'The Case for Simultaneous Toxicity Testing at Different Levels of Biological Organization', in W.E. Bishop, R.D. Cardwell and B.B. Heidolph (eds), *Aquatic Toxicology and Hazard Assessment: Sixth Symposium*, ASTM STP 802, American Society for Testing and Materials, Philadelphia, pp. 111-27

Cairns, J. Jr (1984) 'Multispecies Toxicity Testing', *Environmental Toxicology and Chemistry, 3*, 1-3

Cairns, J. Jr, Heath, A.G. and Parker, B.C. (1975) 'The Effects of Temperature upon the Toxicity of Chemicals to Aquatic Organisms', *Hydrobiologia, 47*, 135-71

Calabrese, E.J. (1983) *Principles of Animal Extrapolation*, Wiley, New York

Cameron, I.L. and Hunter, K.E. (1984) 'Regulation of the Permeability of the Medaka Fish Embryo Chorion to Exogenous Sodium and Calcium Ions', *Journal of Experimental Zoology, 231*, 447-54

Carline, R.F. and Lawal, M.V. (1985) 'Contaminants and Bilateral Asymmetry in Yellow Perch', *Environmental Toxicology and Chemistry, 4*, 543-7

Carson, R. (1962) *Silent Spring*, Fawcett Crest, New York

Casarett, L.J. and Bruce, M.C. (1980) 'Origin and Scope of Toxicology', in J. Doull, C.O. Klaassen and M.O. Amour (eds), *Casarett and Doull's Toxicology: the Basic Science of Poisons*, 2nd edn, Macmillan, New York, Chapter 1

Committee on Methods for Toxicity Tests with Aquatic Organisms (1975)

'Methods for Acute Toxicity Tests with Fish, Macroinvertebrates and Amphibians', US Environmental Protection Agency, Ecological Research Series, EPA-660/3-75-009

Connolly, J.P. (1985) 'Predicting Single-species Toxicity in Natural Water Systems', *Environmental Toxicology and Chemistry, 4*, 573-82

Craik, J.C.A. and Harvey, S.M. (1984) 'Egg Quality in Rainbow Trout: the Relation between Egg Variability, Selected Aspects of Egg Composition, and Time of Stripping', *Aquaculture, 40*, 115-34

Cravens, J.B. and Harrelson, M.E. (1985) 'Thermal Effects', *Journal of the Water Pollution Control Federation, 57*, 649-58

Dabrowski, K. (1984) 'The Feeding of Fish Larvae: Present 'State of the Art' and Perspectives', *Reproduction, Nutrition, Development, 24*, 807-33

Dave, G. (1981) 'Influence of Diet and Starvation on Toxicity of Endrin to Fathead Minnows (*Pimephales promelas*)', EPA-600/S3-81-048, Sept. 1981, United States Environmental Protection Agency

Dave, G. (1984) 'Effect of pH on Pentachlorophenol Toxicity to Embryos and Larvae of Zebrafish (*Brachydanio rerio*)', *Bulletin of Environmental Contamination and Toxicology, 33*, 621-30

Dave, G. (1985) 'The Influence of pH on the Toxicity of Aluminum, Cadmium and Iron to Eggs and Larvae of the Zebrafish, *Brachydanio rerio*', *Ecotoxicology and Environmental Safety, 10*, 253-67

Dave, G., Andersson, K., Berglind, R. and Hasselrot, B. (1981) 'Toxicity of Eight Solvent Extraction Chemicals and of Cadmium to Water Fleas, *Daphnia magna*, Rainbow Trout, *Salmo gairdneri*, and Zebrafish, *Brachydanio rerio*', *Comparative Biochemistry and Physiology, 69C*, 83-98

Davies, R.P. and Dobbs, A.J. (1984) 'The Prediction of Bioconcentration in Fish', *Water Research, 18*, 1253-62

Davis, J.C. (1975) 'Minimal Dissolved Oxygen Requirements of Aquatic Life with Emphasis on Canadian Species: a Review', *Journal of the Fisheries Research Board of Canada, 32*, 2295-332

Dickson, W. (1983) 'Liming Toxicity of Aluminum to Fish', *Vatten, 39*, 400-4

DiMichele, L. and Taylor, M.H. (1981) 'The Mechanism of Hatching in *Fundulus heteroclitus*', *Journal of Experimental Zoology, 217*, 73-9

Dixon, D.G. and Sprague, J.B. (1981a) 'Acclimation-induced Changes in Toxicity of Arsenic and Cyanide to Rainbow Trout', *Journal of Fish Biology, 18*, 579-89

Dixon, D.G. and Sprague, J.B. (1981b) 'Acclimation to Copper by Rainbow Trout (*Salmo gairdneri*) — a Modifying Factor in Toxicity', *Canadian Journal of Fishery and Aquatic Sciences, 38*, 880-8

Dominguez, S.E. and Chapman, G.A. (1984) 'Effect of Pentachlorophenol on the Growth and Mortality of Embryos and Juvenile Steelhead Trout', *Archives of Environmental Contamination and Toxicology, 13*, 739-43

Donaldson, E.M. and Scherer, E. (1983) 'Methods to Test and Assess Effects of Chemicals on Reproduction in Fish', in W.B. Vouk and P.J. Sheehan (eds), *Methods for Assessing the Effects of Chemicals on Reproductive Functions*, Wiley, Toronto, pp. 365-404

Ellehausen, H., Guth, J.A. and Essen, H.B. (1980) 'Factors Determining the Bioaccumulation Potential of Pesticides in the Individual Compartments of Aquatic Food Chains', *Ecotoxicology and Environmental Safety, 4*, 134-57

Falk, D.L. and Dunson, W.A. (1977) 'The Effect of Season and Acute Sublethal Exposure on Survival Times of Brook Trout at low pH', *Water Research, 11*, 13-15

Fish, W. and Morel, F.M.M. (1983) 'Characterization of Organic Copper-complexing Agents Released by *Daphnia magna*', *Canadian Journal of*

190   *Toxicity Testing Procedures*

Fisheries and Aquatic Sciences, 40, 1270-7
Gerking, S.D. (1980) 'Fish Reproduction and Stress', in M.A. Ali (ed.), *Environmental Physiology of Fishes,* Plenum, New York, pp. 569-87
Giles, M.A. and Klaverkamp, J.F. (1982) 'The Acute Toxicity of Vanadium and Copper to Eyed Eggs of Rainbow Trout (*Salmo gairdneri*)', *Water Research, 16,* 885-9
Gillett, J.W. (1983) 'A Comprehensive Prebiological Screen for Ecotoxicological Effects', *Environmental Toxicology and Chemistry, 2,* 463-76
Grande, M. (1967) 'Effect of Copper and Zinc on Salmonid Fishes', *Advances in Water Pollution Research, 3,* 97-111
Grande, M. and Andersen, S. (1983) 'Lethal Effects of Hexavalent Chromium, Lead and Nickel on Young Stages of Atlantic Salmon (*Salmo salar* L.) in Soft Water', *Vatten, 39,* 405-16
Hayton, W.L. and Stehly, G.R. (1983) 'pH Control of Weak Electrolyte Toxicity to Fish', *Environmental Toxicology and Chemistry, 2,* 325-8
Hedtke, S.F. and Arthur, J.W. (1985) 'Evaluation of a Site-specific Water Quality Criterion for Pentachlorophenol Using Outdoor Experimental Streams', in R.D. Cardwell, R. Purdy and R.C. Bahner (eds), *Aquatic Toxicology and Hazard Assessment: Seventh Symposium,* ASTM STP 854, American Society for Testing and Materials, Philadelphia, pp. 551-64
Henriksson, L. and Oscarsson, H.G. (1981) 'Corixids (*Hemiptera — Heteroptera*) the New Top Predators in Acidified Lakes', *Verhandlungen internationale Vereinigung für theoretische und angewandte Limnologie, 21,* 1616-20
Hermens, J., Leeuwangh, P. and Musch, A. (1984) 'Quantitative Structure-Activity Relationships and Mixture Toxicity Studies of Chloro- and Alkylanilines at an Acute Lethal Toxicity Level to the Guppy (*Poecilia reticulata*)', *Ecotoxicology and Environmental Safety, 8,* 388-94
Hodson, P.V. and Sprague, J.B. (1975) 'Temperature-induced Changes in Acute Toxicity of Zinc to Atlantic Salmon (*Salmo salar*)', *Journal of the Fisheries Research Board of Canada, 32,* 1-10
Hodson, P.V., Ross, C.W., Niimi, A.J. and Spry, D.J. (1977) 'Statistical Considerations in Planning Aquatic Bioassays', *Proceedings 3rd Aquatic Toxicity Workshop, Halifax, N.S., 2-3 November, 1976.* Environmental Protection Service Technical Report No. EPS-5-AR-77-1, Halifax, Canada, pp. 15-31
Hodson, P.V., Blunt, B.R. and Whittle, D.M. (1983) 'Suitability of a Biochemical Method for Assessing the Exposure of Fereal Fish to Lead', in W.E. Bishop, R.D. Cardwell and B.B. Heidolph (eds), *Aquatic Toxicology and Hazard Assessment: Sixth Symposium,* ASTM STP 802, American Society for Testing and Materials, Philadelphia, pp. 389-405
How, M.J. (1980) 'The Application and Conduct of Ring Tests in Aquatic Toxicology', *Water Research, 14,* 293-6
Hurlbert, S.H. (1975) 'Secondary Effects of Pesticides on Aquatic Ecosystems', *Residue Reviews, 58,* 81-148
ISO (1982) 'Water Quality Determination of the Inhibition of the Mobility of *Daphnia magna* Straus (Cladocera, Crustacea)', ISO 6341-1982
ISO (1984) 'Water Quality Determination of the Acute Lethal Toxicity of Substances to a Freshwater Fish (*Brachydanio rerio* Hamilton-Buchanan (Teleostei, Cyprinidae)) — Part 1: Static Method, Part 2: Semi-static Method, Part 3: Flow-through Method', ISO 7346 1-3-1984
Karlsson-Norrgren, L., Dickson, W., Ljungberg, O. and Runn, P. (1986) 'Acid Water and Aluminum Exposure; Gill Lesions and Aluminum Accumulation in Farmed Brown Trout (*Salmo trutta*)', *Journal of Fish Diseases,* in press
Kenaga, E.E. (1982) 'Predictability of Chronic Toxicity from Acute Toxicity of

Chemicals in Fish and Aquatic Invertebrates', *Environmental Toxicology and Chemistry*, *1*, 347-58

Kilgour, D.M., McCauley, R.W. and Kwain, W. (1985) 'Modeling the Lethal Effects of High Temperature on Fish', *Canadian Journal of Fisheries and Aquatic Sciences*, *42*, 947-51

Klaverkamp, J.F., Macdonald, W.A., Lillie, W.R. and Lutz, A. (1983) 'Joint Toxicity of Mercury and Selenium in Salmonid Eggs', *Archives of Environmental Contamination and Toxicology*, *12*, 415-19

Könemann, H. (1981) 'Fish Toxicity Tests with Mixtures of more than Two Chemicals: a Proposal for a Quantitative Approach and Experimental Results', *Toxicology*, *19*, 229-38

Könemann, H. and Musch, A. (1981) 'Quantitative Structure Activity Relationships in Fish Toxicity Studies. Part 2: The Influence of pH on the QSAR for Chlorophenols', *Toxicology*, *19*, 223-8

Kovacs, T.G. and Leduc, G. (1982) 'Acute Toxicity of Cyanide to Rainbow Trout (*Salmo gairdneri*) Acclimated at Different Temperatures', *Canadian Journal of Fisheries and Aquatic Sciences*, *39*, 1426-9

Landner, L., Neilson, A.H., Sörensen, L., Tärnholm, A. and Viktor, T. (1985) 'Short-term Test for Predicting the Potential of Xenobiotics to Impair Reproductive Success in Fish', *Ecotoxicology and Environmental Safety*, *9*, 282-93

Landrum, P.F., Reinhold, M.D., Nihart, S.R. and Eadie, B.J. (1985) 'Predicting the Bioavailability of Organic Xenobiotics to *Pontoporeia hoyi* in the Presence of Humic and Fulvic Materials and Natural Dissolved Organic Matter', *Environmental Toxicology and Chemistry*, *4*, 459-67

Larsson, Å., Haux, C. and Sjöbeck, M.-L. (1985) 'Fish Physiology and Metal pollution: Results and Experiences from Laboratory and Field Studies', *Ecotoxicology and Environmental Safety*, *9*, 250-81

LeBlanc, G.A. (1984) 'Interspecies Relationships in Acute Toxicity of Chemicals to Aquatic Organisms', *Environmental Toxicology and Chemistry*, *3*, 47-60

LeBlanc, G.A. and Surprenant, D.C. (1985) 'A Method of Assessing the Toxicity of Contaminated Freshwater Sediments', in *Aquatic Toxicology and Hazard Assessment: Seventh Symposium*, ASTM STP 854, American Society for Testing and Materials, Philadelphia, pp. 269-83

Lewis, M.A. and Perry, R.L. (1981) 'Acute Toxicities of Equimolar and Equitoxic Surfactant Mixtures to *Daphnia magna* and *Lepomis macrochirus*', in D.R. Branson and K.L. Dickson (eds), *Aquatic Toxicology and Hazard Assessment: Forth Conference*, ASTM STP 737, American Society for Testing and Materials, pp. 402-18

Lewis, M.A. and Wee, V.T. (1983) 'Aquatic Safety Assessment for Cationic Surfactants', *Environmental Toxicology and Chemistry*, *2*, 105-18

Lewis, P.A. and Weber, C.I. (1985) 'A Study of the Reliability of *Daphnia* Acute Toxicity Tests', in R.D. Cardwell, R.E. Purdy and R.C. Bahner (eds), *Aquatic Toxicology and Hazard Assessment: Seventh Symposium*, ASTM STP 854, American Society for Testing and Materials, Philadelphia, pp. 73-86

Lipnick, R.L. (1985) 'A Perspective on Quantitative Structure-Activity Relationships in Ecotoxicology', *Environmental Toxicology and Chemistry*, *4*, 255-7

Lloyd, R. (1961) 'Effect of Dissolved Oxygen Concentrations on the Toxicity of Several Poisons to Rainbow Trout (*Salmo gairdneri* Richardson), *Journal of Experimental Biology*, *38*, 447-55

McCarthy, J.F. (1983) 'Role of Particulate Organic Matter in Decreasing Accumulation of Polynuclear Aromatic Hydrocarbons by *Daphnia magna*', *Archives of Environmental Contamination and Toxicology*, *12*, 559-68

McCrady, J.K. and Chapman, G.A. (1979) 'Determination of Copper Complexing Capacity of Natural River Water, Well Water and Artificially Reconstituted Water', *Water Research, 13*, 143-50

McDonald, D. and Wood, C. (1982) 'Branchial and Renal Acid and Ion Fluxes in the Rainbow Trout, *Salmo gairdneri*, at Low Environmental pH', *Journal of Experimental Biology, 93*, 101-18

Macek, K.J. and Sleight, B.H. III (1977) 'Utility of Toxicity Tests with Embryos and Fry of Fish in Evaluating Hazards Associated with the Chronic Toxicity of Chemicals to Fishes', in F.L. Mayer and J.L. Hamelink (eds), *Aquatic Toxicology and Hazard Evaluation*, ASTM STP 634, American Society for Testing and Materials, pp. 137-46

McKim, J.M. (1977) 'Evaluation of Tests with the Early Life Stages of Fish for Predicting Long-term Toxicity', *Journal of the Fisheries Research Board of Canada, 34*, 1148-54

McLeay, D.J. and Munro, J.R. (1979) 'Photoperiodic Acclimation and Circadian Variations in Tolerance of Juvenile Rainbow Trout (*Salmo gairdneri*) to Zinc', *Bulletin of Environmental Contamination and Toxicology, 23*, 552-7

McWilliams, P.G. (1982) 'The Effects of Calcium on Sodium Fluxes in the Brown Trout, *Salmo trutta*, in Neutral and Acid Water', *Journal of Experimental Biology, 96*, 439-42

Maki, A.W. (1983) 'Ecotoxicology — Critical Needs and Creditability', *Environmental Toxicology and Chemistry, 2*, 259-60

Maki, A.W. and Bishop, W.E. (1979) 'Acute Toxicity of Surfactants to *Daphnia magna* and *Daphnia pulex*', *Archives of Environmental Contamination and Toxicology, 8*, 599-612

Manner, H.W. and Muehlman, C. (1975) 'Permeability and Uptake of ^{3}H-uridine during Teleost Embryogenesis', *Science of Biology Journal, 1*, 81-2

Maleug, K.W., Schuytema, G.S., Gakstatter, J.H. and Krawczyk, D.F. (1984) 'Toxicity of Sediments from Three Metal Contaminated Areas', *Environmental Toxicology and Chemistry, 3*, 279-91

Marking, L.L. (1975) 'Effects of pH on Toxicity of Antimycin on Fish', *Journal of the Fisheries Research Board of Canada, 32*, 769-73

Marking, L.L. and Dawson, V.K. (1973) 'Toxicity of Quinaldine Sulfate to Fish', *Investigations in Fish Control No. 48*, U.S. Department of the Interior, Fish and Wildlife Service

Marking, L.L. and Olson, L.E. (1975) 'Toxicity of the Lampricide 3-Trifluoromethyl-4-nitrophenol (TFM) to Nontarget Fish in Static Tests', *Investigations in Fish Control No. 60*, U.S. Department of the Interior, Fish and Wildlife Service

Marking, L.L., Bills, T.D. and Crowther, J.R. (1984) 'Effects of Five Diets on Sensitivity of Rainbow Trout to Eleven Chemicals', *Progressive Fish-Culturist, 46*, 1-5

Mayer, F.L. (1983) 'Clinical Tests in Aquatic Toxicology: a Paradox?', *Environmental Toxicology and Chemistry, 2*, 139-40

Mehrle, P.M., Mayer, F.L. and Johnson, W.W. (1977) 'Diet Quality in Fish Toxicology: Effects on Acute and Chronic Toxicity', in F.L. Mayer and J.L. Hamelink (eds), *Aquatic Toxicology and Hazard Evaluation*, ASTM STP 634, American Society for Testing and Materials, pp. 269-80

Mount, D.I. (1977) 'Present Approaches to Toxicity Testing — a Perspective', in F.L. Mayer and J.L. Hamelink (eds), *Aquatic Toxicology and Hazard Evaluation*, ASTM STP 634, American Society for Testing and Materials, pp. 5-14

Nebeker, A.V., Cairns, M.A., Gakstatter, J.H., Maleug, K.W., Schuytema, G.S. and Krawczyk, D.F. (1984) 'Biological Methods for Determining Toxicity of

Contaminated Freshwater Sediments to Invertebrates', *Environmental Toxicology and Chemistry, 3*, 617-30

Niblett, P.D. and McKeown, B.A. (1980) 'Effect of the Lamprey Larvicide TFM (3-trifluoromethyl-4-nitrophenol) on Embryonic Development of the Rainbow Trout (*Salmo gairdneri*, Richardson)', *Water Research, 14*, 515-19

Norberg, T.J. and Mount, D.I. (1985) 'A New Fathead Minnow (*Pimephales promelas*) Subchronic Toxicity Test', *Environmental Toxicology and Chemistry, 4*, 711-18

Norton, M.G. and Franklin, F.L. (1980) 'Research into Toxicity Evaluation and Control Criteria of Oil Dispersants', Fisheries Research Technical Report No. 57, Ministry of Agriculture, Fisheries and Food, Directorate of Fisheries Research, Lowestoft, 20 pp.

Ohr, E. (1976) 'Tricaine Methanesulfonate — I. pH and its Effects on Anaesthetic Potency', *Comparative Biochemistry and Physiology, 54C*, 13-17

O'Neil, R.V., Bartell, S.M. and Gardner, R.H. (1983) 'Patterns of Toxicological Effects in Ecosystems: a Modelling Study', *Environmental Toxicology and Chemistry, 2*, 451-61

Peakall, D.B. and Fox, G.A. (1983) 'Evaluation of Effects from Ecotoxicological Tests', in K. Christiansen, B. Koch and F. Bro-Rasmussen (eds), *Chemicals in the Environment, Proceedings of the International Symposium, Lyngby-Copenhagen-Denmark, 18-20 October 1982*, The Technical University of Denmark, Lyngby, Denmark, pp. 267-83

Peterson, R.H. (1976) 'Temperature Selection of Juvenile Atlantic Salmon (*Salmo salar*) as Influenced by Various Toxic Substances', *Journal of the Fisheries Research Board of Canada, 33*, 1722-30

Prater, B.L. and Andersson, M.A. (1977) 'A 96-hour Sediment Bioassay of Duluth and Superior Harbor Basins (Minnesota) Using *Hexagenia limbata, Asellus communis, Daphnia magna*, and *Pimephales promelas* as Test Organisms', *Bulletin of Environmental Contamination and Toxicology, 18*, 159-69

Raleigh, R.F., Hickman, T., Solomonn, R.C. and Nelson, P.C. (1984) 'Habitat Suitability Information: Rainbow Trout', US Fish and Wildlife Service, FWS/OBS-82/10. 60, 64 pp.

Rask, M. (1984) 'The Effect of Low pH on Perch, *Perca fluviatilis* L. II. The Effect of Acid Stress on Different Developmental Stages of Perch', *Annales Zoologici Fennici, 21*, 9-13

Rosenthal, H. and Alderdice, D.F. (1976) 'Sublethal Effects of Environmental Stressors, Natural and Pollutional, on Marine Fish Eggs and Larvae', *Journal of the Fisheries Research Board of Canada, 33*, 2047-65

Russo, R.C., Thurston, R.V. and Emerson, K. (1981) 'Acute Toxicity of Nitrite to Rainbow Trout (*Salmo gairdneri*): Effects of pH, Nitrite Species, and Anion Species', *Canadian Journal of Fisheries and Aquatic Sciences, 38*, 387-93

Scherer, E. (ed.) (1979) 'Toxicity Tests for Freshwater Organisms', *Canadian Special Publications of Fisheries and Aquatic Sciences, 44*, 194 pp.

Spehar, R.L. (1976) 'Cadmium and Zinc Toxicity to Flagfish, *Jordanella floridae*', *Journal of the Fisheries Research Board of Canada, 33*, 1939-45

Spehar, R.L. and Carlson, A.R. (1984) 'Determination of Site-specific Water Quality Criteria for Cadmium and the St. Louis River Basin, Duluth, Minnesota', *Environmental Toxicology and Chemistry, 3*, 651-65

Spehar, R.L., Nelson, H.P., Swanson, M.J. and Renoos, J.W. (1985) 'Pentachlorophenol Toxicity to Amphipods and Fathead Minnows at Different pH Values', *Environmental Toxicology and Chemistry, 4*, 389-97

Sprague, J.B. (1970) 'Measurement of Pollutant Toxicity to Fish. 2. Utilizing and Applying Bioassay Results', *Water Research, 4*, 3-22

Sprague, J.B. (1976) 'Current Status of Sublethal Tests of Pollutants on Aquatic Organisms', *Journal of the Fisheries Research Board of Canada, 33*, 1988-92

Stebbing, A.R.D. (1981) 'Hormesis-stimulation of Colony Growth in *Campanularia flexuosa (Hydrozoa)* by Copper, Cadmium and Other Toxicants', *Aquatic Toxicology, 1*, 227-38

Stebbing, A.R.D. (1982) 'Hormesis — the Stimulation of Growth by Low Levels of Inhibitors', *Science of the Total Environment, 22*, 213-34

Stephan, C.E. (1977) 'Methods for Calculating an LC50 ', in F.L. Mayer and J.L. Hamelink (eds), *Aquatic Toxicology and Hazard Evaluation*, ASTM STP 634, American Society for Testing and Materials, pp. 65-84

Stephan, C.E. (1985) 'Are the "Guidelines for Deriving Numerical National Water Quality Criteria for the Protection of Aquatic Life and its Uses" Based on Sound Judgements?', in R.D. Cardwell, R. Purdy and R.C. Bahner (eds), *Aquatic Toxicology and Hazard Assessment: Seventh Symposium*, ASTM STP 854, American Society for Testing and Materials, Philadelphia, pp. 515-26

Stevens, D.G. and Chapman, G.A. (1984) 'Toxicity of Trivalent Chromium to Early Life Stages of Steelhead Trout', *Environmental Toxicology and Chemistry, 3*, 125-33

Suter, G.W. III, Vaughan, D.S. and Gardner, R.H. (1983) 'Risk Assessment by Analysis of Extrapolation Error: a Demonstration for Effects of Pollutants on Fish', *Environmental Toxicology and Chemistry, 2*, 369-78

Takimoto, Y., Hagino, S., Yamada, H. and Miyamoto, J. (1984) 'The Acute Toxicity of Fenitrothion to Killifish (*Oryzias latipes*) at Twelve Different Stages of its Life History', *Journal of Pesticide Science, 9*, 463-70

Tarifeño-Silva, E., Kawasaki, L.Y., Yu, D.P., Gordon, M.S. and Chapman, D.J. (1982) 'Aquacultural Approaches to Recycling of Dissolved Nutrients in Secondary Treated Domestic Wastewaters — III. Uptake of Dissolved Heavy Metals by Artificial Food Chains', *Water Research, 16*, 59-65

Tatem, H.E., Anderson, J.W. and Neff, J.M. (1976) 'Seasonal and Laboratory Variations in the Health of Grass Shrimp *Palaemonetes pugio*: Dodecyl Sodium Sulphate Bioassay', *Bulletin of Environmental Contamination and Toxicology, 16*, 368-75

Taylor, M.J. (1985) 'Effect of Diet on the Sensitivity of *Daphnia magna* to Linear Alkylbenzene Sulfonate', in R.D. Cardwell, R. Purdy and R.C. Bahner (eds), *Aquatic Toxicology and Hazard Assessment: Seventh Symposium*, ASTM STP 854, American Society for Testing and Materials, Philadelphia, pp. 53-72

Theilacker, G. and Dorsey, K. (1980) 'Larval Fish Diversity, a Summary of Laboratory and Field Research', in G. Sharp (rapporteur), *Workshop on the Effects of Environmental Variation on the Survival of Larval Pelagic Fishes, UNESCO, Intergovernmental Oceanographic Commission, Workshop Report No. 28*, 105-42

Tovell, P.W.A., Newsome, C. and Howes, D. (1974) 'Effects of Water Hardness on the Toxicity of an Anionic Detergent to Fish', *Water Research, 8*, 291-6

Van den Berg, C.M.G., Wong, P.T.S. and Chau, Y.K. (1979) 'Measurement of Complexing Materials Excreted from Algae and their Ability to Ameliorate Copper Toxicity', *Journal of the Fisheries Research Board of Canada, 36*, 901-5

Van Hoof, F., Van Craenenbroeck, W.V. and Marivoet, D. (1984) 'Investigations into the Causes of Fish Kills Occuring in the River Meuse', in D. Pascoe and R.W. Edwards (eds), *Freshwater Biological Monitoring*, Pergamon, Oxford, pp. 53-63

Warren, C.E. (1971) *Biology and Water Pollution Control*, Saunders, Philadelphia

Wickins, J.F. (1981) 'Water Quality Requirements for Intensive Aquaculture: a Review', in *Proceedings from the World Symposium on Aquaculture in Heated*

Effluents and Recirculation Systems, Stavanger 28-30 May, 1980, vol. 1, Berlin, pp. 17-37

Woltering, D.M. (1984) 'The Growth Response in Fish Chronic and Early Life Stage Toxicity Tests: a Critical Review', *Aquatic Toxicology*, 5, 1-21

Zitko, V. and Carson, W.G. (1977) 'Seasonal and Developmental Variation in the Lethality of Zinc to Juvenile Atlantic Salmon (*Salmo salar*)', *Journal of the Fisheries Research Board of Canada*, 34, 139-41

INDEX

abductor muscles 86, 92
Abiotic factors in toxicology 175
acetylcholine 128
acid-base regulation 24-47, 51
    ion transfer in 28-9, 51
adductor muscles 86
adrenaline 64, 65, 96, 112
adrenergic nerves 90, 91
alimentary tract see gut
aluminiun 173
δ-aminolevulinic acid (ALA) 11
δ-aminolevulinic acid dehydratase
    (ALA-D) 161
ammonia 30
antigen-trapping cells 10
appetite 140
atrium 71, 74
autonomic nerves
    of the heart 110
    of the gills 90, 92
    of the gut 121, 126, 140

bicarbonate 33, 40, 41, 42, 43, 44,
    45, 53
biotic factors in toxicology 180
blood
    haemopoiesis in 4
    of marlin 51-7
blood cells
    maturation stages 3
bombesin 122, 128, 129, 130, 133
branchial arteries 63, 87
branchial nerves 89, 91
branchial veins 87
buffering 24-6
    capability 25
    capacity 25, 26, 60
bulbus arteriosus 72, 76
bulbus cordis 72

Ca^{2+}
    in myocardial cells 78-80
cadmium 159, 162, 163, 164, 173
caerulein 123, 132
cardiac cycles 72-7
catecholamines 64, 96, 110, 113, 114
chlorinated hydrocarbons 159
cholecystokinin (CCK) 123, 132

chromium 173
colloid osmotic pressure 54, 153
conus arteriosus 72
coronary vasculature 77
    of marlin 63-4

DDT 179
digestion 130
'Dunel-Laurent sphincters' 90-4

EC50 171
eledoisin 134, 135
embryonic development
    effects of pollutants 172-4
endocrine cells 119, 120-6
endorphin 123
energy metabolism
    in myocardium 65, 81-3
    in somatic muscle 58-62
    in toxicology 162-6
enkephalin 123, 129, 130
enteric nervous system 126, 140
epigonal organ 9, 10
erythrocytes 1, 3
erythropoiesis 16, 162
erythropoietic tissues 4
erythropoietin 16
exercise 53, 102-15
    and plasma catecholamines
        112-15
    cardiovascular control in 110-12
    circulatory adjustments in 106-10
    respiratory adjustments in 106-10

ferritin 12
food processing rate 142-51
food reception 140

gastric emptying 142-51
gastric secretion 132-5
gastrin 123, 132
gastro-intestinal peptides 119-35
gills 86-99
    muscles 86
    vasculature: anatomy 87-9; flow
        patterns 97-9; humoral control
        96-7; innervation 89-94
glomerular filtration 153-6

glucagon 124
granulocytes 1, 5
granulopoietic tissues 5-7
gut 119-51
    endocrine cells 120-6
    food processing rate 142-51
    food reception 140
    motility 128-31
    nerves 126-7

haemoglobin 3, 51, 52, 161, 163
    synthesis 11
haemopoiesis 1, 4, 17
    hormonal control 16
haemopoietic tissue 3-11
haemosiderin 12
hatching 172
head kidney 5
heart 71-83, 87, 109
    electromechanical coupling 78-80
    of marlin 63-7
    structure 71-2
histamine 132, 134
hormesis 173
5-hydroxytryptamine (5-HT) 129
hypercapnia 33-6, 40-7
hyperventilation 27

immunoglobulins 10
immunoreactivity 120
insulin 124
intestine *see* gut
ion transfer 28, 35
    effects of pollutants 164-5

kidney 153-6

lactacidosis 31-3, 37-40
lactate 39, 51, 52, 163
lactic acid 31, 37
lamellar recruitment 97-8
LC50 171, 178
lead 159, 162, 173
leucocytes 1, 5, 163
Leydig organ 6, 8, 15
lymphatic tissues 3
lymphocytes 1, 2
lymphohaemopoietic tissues 3
lympoid tissues 3
lymphomyeloid complex 4
lymphomyeloid tissues 3, 12, 19

macrophages 10
melanomacrophage centres 12

mercury 173
metallothionein 163, 164
mixed function oxidase (MFO) 161,
    165-6
monocytes 1
mortality 171
muscle
    somatic locomotor 37, 57-61, 65,
        102-5, 109
    visceral: gills 86, 92; gut 127-40
myenteric plexus 126
myocardium 63, 65, 66
    Ca^{2+} in myocardial cells 78-80
    morphology of 77-8

neurotensin 124, 129, 130

oxygen uptake 36, 107, 108, 109
    in marlin 51, 56

pancreas 124, 163
pepsin 133
peptides 119-35
pericardial tissue 7, 9
pericardium 72, 73
physalaemin 135
plasma cells 1, 10
pollution 158-87
porphyrin 12
pyloric sphincter 145, 147

ram ventilation 107, 108
rectum *see* gut
reproduction
    effects of pollutants 174
reticuloendothelial tissues 3

sinus venosus 71, 74
somatostatin 122, 125, 129, 134, 141
spindle cells 1
spleen 4, 14, 15
stem cells 2
stomach *see* gut
stress 96, 113, 162
submucous plexus 127
substance P 122, 125, 129, 130
swimming performance 105-6
    in marlin 53-7

tachykinins 126, 135
thrombocytes 1
toxicity 170-87
    abiotic factors: alkalinity,
        hardness, pH 176; dissolved

organics 177; dissolved oxygen
177; salinity 177-8; suspended
solids 178-9; temperature
179-80
biotic factors: developmental stage
180-1; diet and nutrition
181-2; previous exposure
182; species 180
mortality 171-2
of effluents 183-4
of mixtures 182-3
on development and growth 172-4
on reproduction 174-5
seasonal variation 184-5
water-quality criteria 185-7
toxicology 158-87

methods in 160-1
sublethal effects 158-66
toxic unit 183
transferrin 12

vasoactive intestinal polypeptide
(VIP) 126, 129, 131, 141
ventral aorta 72, 87
ventricle 71

water quality criteria 184-5
site specific 184
species specific 185
Windkessel effect 72

zinc 173